Statistics: the now and the why

AN INTRODUCTORY COURSE

E H Lockwood
formerly Scholar of St John's College, Cambridge,
and Senior Mathematics Master, Felsted School

John Murray 50 Albemarle Street London

© E H Lockwood 1969

All Rights Reserved. No part of this publication may be reproduced, stored in a retrieval system, or transmitted, in any form or by any means, electronic, mechanical, photocopying, recording or otherwise, without the prior permission of the Copyright owner.

Printed in Great Britain by
William Clowes and Sons, Limited
London and Beccles

7195 1876 8

Contents

	Preface	vii
1	Introduction	1
2	Frequency distributions	4
3	Lines of closest fit. Correlation	26
4	Probability	38
5	Probability distributions	56
6	The binomial distribution	71
7	The Poisson distribution	81
8	The normal distribution	91
9	Sampling	115
	Notation and formulae	138
	Answers	142
	Index	149

CORRIGENDA

p. 28, last line. For 4 698 *read* 4 689
p. 40, line 14. For of A given B *read* of B given A
p. 51, Ex. 41. Add a footnote: Lord Yarborough laid 1000 to 1 against its happening.
p. 121, last two formulae. For $+2$ *read* -2
p. 124, 5 (ii). For more than 0·10 kg *read* less than 0·01 kg
p. 129, last formula. For $\frac{1}{t}$ *read* $-\frac{1}{t}$
p. 134, line 8, to read:

$$\text{S.E. of difference} = \sqrt{\left[0{\cdot}583 \times 0{\cdot}417 \left(\frac{1}{150} + \frac{1}{200}\right)\right]}$$
$$= 0{\cdot}05326.$$

—— *line 11. For* $\frac{0{\cdot}04}{0{\cdot}0262} = 1{\cdot}53$ approx. *read* $\frac{0{\cdot}04}{0{\cdot}05326} = 0{\cdot}751$ approx.

—— *line 14. For* 0·126, or about 1 in 8 *read* 0·45 approx.

p. 135. Add a final line: $\therefore$ variance of sample mean $= \frac{N-n}{N-1} \cdot \frac{\sigma^2}{n}$.

p. 142, Ex. 2, 2. For $y = 0{\cdot}407 - 21{\cdot}8$ *read* $y = 0{\cdot}407x - 21{\cdot}8$
p. 143, Ex. 3, 22 (i). For 4850 *read* 19 380

22 (*iv*). *For* $\frac{32}{625}$ *read* $\frac{512}{9025}$

26. *For* $\frac{1}{27}, \frac{26}{27^5}$ *read* (i) $\frac{5}{54}$, (ii) $\left(\frac{5}{54}\right)^4 \left(\frac{49}{54}\right)$

p. 144, Ex. 5, 6. Add: 30·1, 28·7 oz.
12. *For* 1·31 *read* 1·13

—— Ex. 7, *1 (ii). For* $\frac{11}{2048}$ *read* $\frac{61}{1024}$

1 (iii). For $\frac{4089}{4096}$ *read* $\frac{4077}{4096}$

2. *For* 0·358, 0·55 *read* 0·60, 0·915

p. 145, Ex. 7, 11. For 2·4 *read* 2·24
—— Ex. 9, 9. *For* 10 *read* 9
—— Ex. 10, 7 (*ii*). *For* 0·006 *read* 0·6 per cent.
p. 146, Ex. 12, 5 (i). For 0·30 *read* 0·20
—— Ex. 12, 13 (*i*) (*a*) *For* 7·35 *read* 7·34
               (*b*) *For* 5·11, 1·11 *read* 5·21, 1·10
            (*ii*) *For* over 3 times S.D. *read* nearly 10 times S.E.
           (*iii*) *For* 1·36 × S.D. *read* 4 × S.E.
              For Not convincing *etc. read* Highly significant.
           (*iv*) *Insert* (*a*) *before* S.D.

---

0 7195 1876 8 (Cased)
0 7195 1912 8 (Paper)

Lockwood: *Statistics: the how and the why*

Preface

This book is written in the belief that students of the better sort are not content to be given a rule from which practical results can be obtained, and told that 'the proof is beyond the scope of the present volume'. They want to know the reason why.

The proofs given are intended to satisfy this entirely admirable curiosity, and the methods chosen are those that students at sixth form level can most easily understand. Even those readers who omit the starred sections may like to know that these proofs are not entirely inaccessible.

While written primarily for sixth forms, the book may also provide an introduction to the subject for first-year students of science and engineering. The material is selected as being appropriate to an introductory course, and in bulk not too great to be covered in one year, at a few hours a week. Many readers will already have some knowledge of descriptive statistics, but a brief account of that side of the subject is given in Chapters 2 and 3. Thereafter the book is mainly concerned with questions of probability. It is written more for scientists than for students of economics, though questions of sampling will be of interest to both classes of reader.

No methods have been introduced without justification. This requirement has imposed limits on the scope of the book, and some students will find that it does not go far enough for their purposes. The χ^2 test, the t-test for small samples, and the variance ratio test have not been included, because their justification would require a considerable amount of further theory. But the student who has read this book should be in a good position to learn the use of these tests, as explained in books on statistical method, and perhaps to study the theory in more advanced texts.

Some apology may be needed for the use, in certain examples, of such old-fashioned units as the foot, the mile and the ton. No doubt the Metric System will soon be fully established in Great Britain, even for everyday purposes, and the Système Internationale for scientific work, but it seemed proper that quoted figures should be left in their original form.

My thanks are due to the Oxford and Cambridge Joint Board and to the Southern Universities Joint Board for permission to use some of their examination questions.

E. H. L.

Note

* An asterisk is used to indicate sections of the work some readers may prefer to omit.

1
Introduction

The modern science of statistics is born of the union of two quite independent lines of thought: the description of states (in the political sense of the word), from which the name *statistics* is derived; and the theory of probability, which originated from the study of games of chance.

It is said that Aristotle was the first to consider states in comparison with one another. In Europe a considerable literature of this kind grew up in the sixteenth, seventeenth, and eighteenth centuries, the object being to compare the power of the different states. Such studies flourished particularly in Germany, under the name *Staatenkunde*. To the modern mind it seems obvious that any such comparisons must rest on numerical data, but, in fact, they were political and geographical and not at all numerical. They were concerned with the organization of the states, their peoples and governments, their armed forces, commerce and industry, but in general terms only. Numerical facts were simply not available. Estimates of population were vague in the extreme, and other numerical data almost entirely lacking. It is, however, of interest that at one stage a tabular form of comparison was introduced, under the name *Tabellenstatistik*, facts for different countries being placed in parallel columns, a foreshadowing of things to come. But this system was criticized as being restricted to the dry bones, allowing no room for any account of the spirit and genius of a people.

In 1662 a London merchant, John Graunt, published a book called *Natural and Political Observations upon the Bills of Mortality*. This was the beginning of numerical statistics, known at first as 'political arithmetic'. The term was coined by Sir William Petty, a friend of Graunt and an original member of the Royal Society. In his *Five Essays in Political Arithmetick* (1687) Petty attempted to estimate the population of London, using three different methods, each somewhat rough and ready, but agreeing fairly well with one another. (He assumed an annual death rate of 1 in 30, and in plague years 1 in 5.) He concluded that in 1685 the population of London was about 696,000.

Petty was followed by Gregory King, who, in 1696, gave estimates not only of the total population of England, but the distribution by age, class, and income. He gave the total population as 5,500,520, divided into 1,349,586 families; beginning with the 160 families of 'temporal lords', with 40 heads per family (this included indoor servants) and an annual income of £3200, continuing through the social classes down to 400,000 families of 'cottagers and paupers' with an annual income of £6 10*s.* per family of $3\frac{1}{4}$ heads. His distribution by age showed only 2,400,000 above the age of 25, and 600,000 above the age of 60. In spite of the roughness of many of his estimates his work is of the greatest value to social historians.

Quite independently of this 'political arithmetic' there grew up the theory of probability, stimulated by the widespread interest in games of chance. Cardan, a gambler himself, wrote a small paper, *De Ludo Aleae*, discussing the chance of throwing a particular number with two or three dice. He also considered, unsuccessfully, the division of stakes between two players in an unfinished game. This question was the subject of a famous correspondence between Pascal and Fermat, from which the theory of probability is commonly dated.

It was Halley who first brought together these two streams of thought in a union that was to prove so fruitful in the modern theory of statistics. He wrote, in 1693, a paper for the Royal Society (*Phil. Trans.*, **196**, p. 596) entitled, *An Estimate of the Degrees of Mortality of Mankind drawn from curious Tables of the Births and Funerals at the City of Breslaw; with an attempt to ascertain the Price of Annuities upon Lives.* This included a life table, from which he deduced the odds against a person's dying within a year, the price of insuring a life for a stated period, and the price of a life annuity. He also studied chances depending on two, or three, lives. Life assurance offices began to be founded towards the end of the seventeenth century, but it was not till the foundation of the 'Old Equitable', in 1762, that premiums were worked out on a rational basis, on the principles laid down by Halley.

The theory of probability was greatly advanced by Jakob Bernoulli (1654–1705), who considered the problem of repeated trials and showed ('Bernoulli's Theorem') that, as the number of trials is increased, there is a greater probability of getting within a certain percentage of the expected number of successes. The idea was taken up by de Moivre (1667–1754) who, in 1733, found the ratio of the middle term of a binomial expansion to the sum of all the terms, and simplified it by means of the formula of his friend James Stirling. He went on to find the size of the terms at a given distance from the middle term, and the sum of all terms between given limits. In short, he established the formula for the 'Normal Curve of Errors' (commonly attributed to Gauss, who lived a hundred years later, 1777–1855) and showed how to find the area under any part of the curve. He was aware, too, that this work had a bearing on statistical observations and the deduction of laws from them.

Closely allied was the problem of observational errors in astronomy. Given a set of readings that are not in perfect agreement, most people regard the arithmetic mean as the best possible estimate of the true value. This mean has the property that the sum of the squares of the deviations from it is as small as possible, a property that can be applied to the more extended problem of finding a law to fit observed values of a variable. This *method of least squares* was first given by Legendre in 1805.

The comparison of observed results with those to be expected according to some law or theory is the essence of modern statistical work. There is, first, the collection and presentation of data, and the summarizing of them by means of such representative measures as the mean. (The word 'statistic' is now used in the singular for any such representative measure.) Then, the theory of probability is brought in to decide whether the statistics may be regarded as in accordance with some theory or supposition. The ultimate question a statistician must usually answer is: 'Do these results show a significant divergence from what might be expected under the theory?'

2
Frequency distributions

The presentation of data
There are many ways of presenting numerical data and of representing or misrepresenting them diagrammatically. Tabular forms are useful and necessary, but pages thick with figures are repellent to the eye and mind; so diagrams are frequently used as a means of conveying information in a striking manner, making the least possible demand on the reader. This has its dangers, because a lazy reader will not enquire about the exact meaning and origin of the data represented and may well draw entirely false conclusions as a result. For this reason, the proper labelling of diagrams, the marking of scales, and full information about how the figures illustrated have been compiled is of the utmost importance.

'Statistics can be useful, boring, or dangerous. Mostly they are dangerous' (*The Economist*, 25 September 1954). This was part of a comment on a publication giving estimates of the national product of twenty different countries. It was pointed out that the totals at the feet of parallel columns were not necessarily comparable, as they had been converted to a common unit (dollars) by the simple but deceptive method of applying the official exchange rates; and that percentage changes might not be meaningful, because they were based on arbitrary starting-points that might have been years of great prosperity in some countries, and of depression in others. On the same page, perhaps by design, a more striking example was given of the traps into which the reader of statistics might fall:

> '... the marriage rate in Ireland was stated to be only about one-third of that in the United Kingdom. Owing to the fact that the Irish statistics record numbers of marriages and the British ones the number of persons married, the ratio is in fact about two-thirds.'†

† From *The Economist*, by courtesy of the Editor.

'Compare like with like' may seem an obvious dictum, but its application demands a critical attitude on the part of the reader or listener as well as care and honesty on the part of the writer or speaker. Too often figures are published without any indication of their origin or even of their exact meaning. When diagrams are used the same faults are even more prevalent; moreover, the scales chosen may be unsuitable, or altogether lacking. (This applies particularly to pictorial diagrams.)

A false impression can be given by figures that are incomplete, the selection being made in a consciously or unconsciously biased way. A time series may stop at a favourable moment or may start at a peak year. (This last is almost standard procedure in the presentation of claims for increased wages, salaries, grants, or subsidies.)

Sampling procedure requires particular care, as it is only too easy to produce figures with a built-in bias. The classic example was in one of the earlier public opinion polls in the USA, in which a forecast of a Republican victory was afterwards falsified, the error being due to the fact that the sample was based on names taken from such sources as the telephone directories.

An investigation into the length of life of razor-blades was carried out by asking users when they last put in a new blade, the idea being that this would determine the average half-life of a blade. But it did not allow for the fact that the longer-lasting blades had a better chance of being chosen. For the same reason, an infant's expectation of life is not equal to twice the average age of the existing population. (In fact, it is much less. *See* p. 55, No. 10.)

Figures of occupational mortality may likewise be deceptive. Thus, the average age at death for innkeepers would doubtless be higher than that recorded for professional footballers, but it is not therefore a more healthy occupation.

Averages

An average is a representative measure, an attempt to summarize in one figure a considerable body of information. Naturally, it does not tell the whole story. What is gained by disregarding minor fluctuations may be balanced by the loss from the omission of relevant facts. A bowler who appears in one match and takes 2 wickets for 20 runs may have the same average as one who bowls all through the season taking 40 wickets for 400 runs, but their performances are not equal. If they both appear in one more match and both take one wicket for 40 runs, their averages for the season will differ widely. Similar to this are the well-known puzzles of which Exercise 1, No. 4, is an example.

The 'average of the averages' is not the same as 'the average'. This is obvious enough in simple cases. For example, if 4 boys of average age 16 join a class of 20, of which the average age is 17, no one would suppose that the new average would be $16\frac{1}{2}$. To obtain the correct result the figures must be 'weighted', i.e.

$$\text{new average age} = \frac{4 \times 16 + 20 \times 17}{4 + 20}, = \frac{404}{24}, = 16 \cdot 83 \text{ approx.}$$

Weighted averages are used for many purposes, notably in the calculation of index numbers. The 'retail price of food' index must clearly depend not only on the prices of the different foods but also on the quantities supposed to be consumed. An increase in the price of caviar should not affect the index as much as a similar increase in that of bread or ice-cream.

The average gives less information than the original figures, as the cricket example showed. To compare the (crude) death rates of two towns may be quite misleading if the populations have different age distributions. Suppose, for example, that the figures are as follows:

Age	No. of people		No. of deaths		Deaths per 1000	
	Town A	Town B	A	B	A	B
Over 60	8,000	6,000	80	63	10	10·5
40–60	10,000	8,000	150	124	15	15·5
20–40	12,000	14,000	84	100	7	7·1
0–20	10,000	12,000	100	125	10	10·4
Total	40,000	40,000	414	412	10·35	10·30

It will be seen that the crude death rates for the two towns are nearly equal, although town A has a lower death rate for each of the age groups. This is because town B has a younger population than town A. To make a fair comparison it is better to calculate a 'standardized death rate', compiled from the death-rates of the age groups by weighting them according to an imaginary population with a standard age distribution.

Diagrams

Many different forms of diagram are used for displaying numerical data, the basic types being as follows:

(i) the 'time chart', an ordinary graph with time as the independent variable (sometimes with a logarithmic scale for the dependent variable) (Fig. 2.1);

Frequency distributions

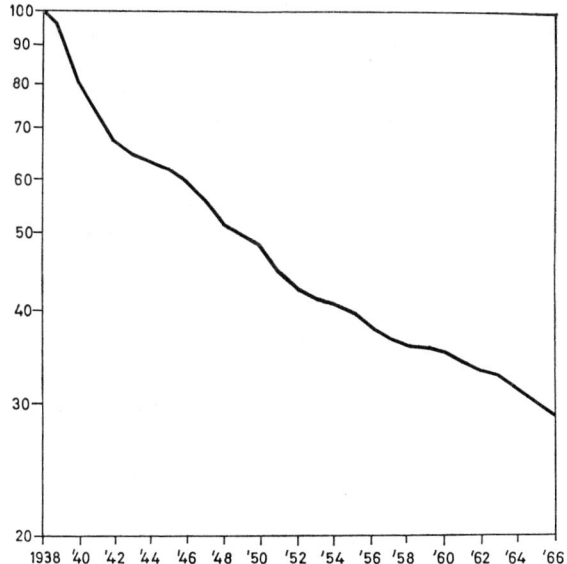

Fig. 2.1. Fall in purchasing value of the £, based on annual values of the Consumer Price Index
(Log scale to show relative fall)
(Reproduced by permission of Times Newspapers Ltd.)

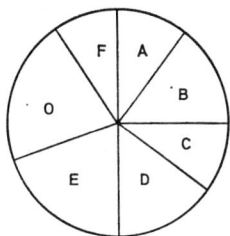

Fig. 2.2. Number of candidates in grades in an A-level examination

(ii) the 'pie diagram', for illustrating a proportional division (Fig. 2.2);
(iii) the 'bar diagram', for showing the numbers in different categories (Fig. 2.3);
(iv) the 'frequency diagram', for showing how frequently the different values of a variable occur (Figs 2.4 and 2.5).

8   Statistics: the how and the why

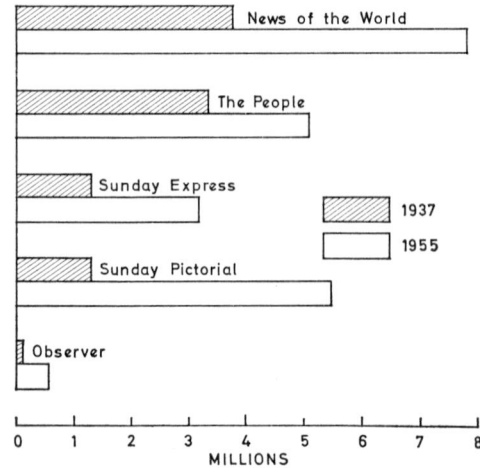

Fig. 2.3. Circulation of Sunday newspapers 1937–55

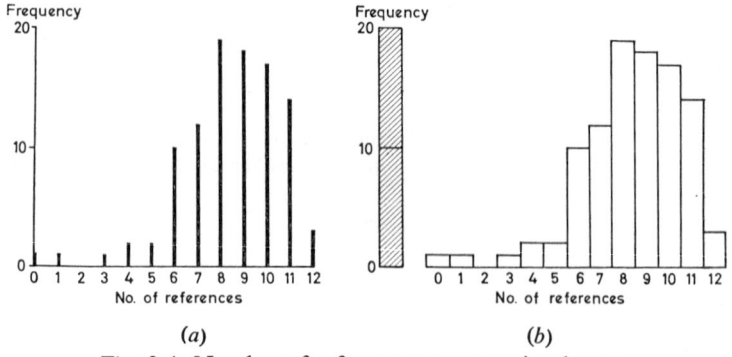

Fig. 2.4. Number of references per page in glossary

Variables

A *variable* is a letter, such as x, that denotes any one of some particular set of values, finite or infinite. In (i) above, the term 'independent variable' is used in the ordinary mathematical sense of a variable for which arbitrary values may be chosen; but in (iv) the values are those that are determined, or might be determined, by a series of observations. Such a variable is called a *random variable*. A *frequency distribution* is a record of the number of times each value, or group of values, occurs in such a series.

There is, moreover, a distinction between *discrete variables*, for

Frequency distributions

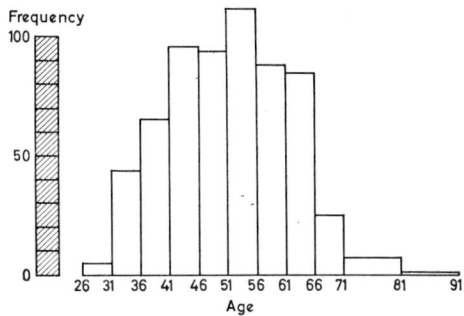

Fig. 2.5. Age distribution of MPs, 1964

which the possible values are separated by finite intervals, and *continuous variables*, which can take any values within certain ranges.

Example 2.1

As an example of a discrete distribution, the following table shows the number of references per page in a glossary of sea and naval terms in Shakespeare:

No. of references	Frequency (i.e. the number of pages on which the stated number of references occurred)
0	1
1	1
2	0
3	1
4	2
5	2
6	10
7	12
8	19
9	18
10	17
11	14
12	3
	100

The frequencies may be represented by a series of 'vertical' lines (Fig. 2.4a), or the lines may be widened into columns (Fig. 2.4b), in this case purely for display.

Example 2.2

For a continuous distribution, columns serve a useful purpose, their width representing the spread or 'width' of the group, and their areas the frequency for that group. For example, the age distribution of members of the House of Commons after the 1964 election was as follows:

Age	Frequency
81+	1
71–80	14
66–70	25
61–65	85
56–60	88
51–55	112
46–50	94
41–45	96
36–40	66
31–35	43
26–30	5
	629

The distribution is illustrated in Fig. 2.5. Diagrams of this type are called *histograms*.

Care must be taken in labelling the boundaries of the groups. Assuming that 'age' means 'age last birthday', the boundaries in this case are 31, 36, etc. The groups, in fact, cover the ages 26–31, 31–36, etc. It is often convenient to specify them by the central values, in this case $28\frac{1}{2}$, $33\frac{1}{2}$, etc. The fact that frequency is represented by area is important for the group 71–81, where the 'width' of the group is 10 years, not 5, and thus the height of the column is half what might have been expected for a linear scale. For the group 81+, an arbitrary length of base must be assigned.

The mode

The diagrams for both the above examples show a rise to a maximum and then a falling away. The value of the variable corresponding to the maximum is called the *mode*. In Example 2.1 the mode is 8, and in Example 2.2, where it is difficult to decide on an exact maximum point, we may say that the modal group is 51–55. This type of distribution is the commonest and is called *unimodal*. There is a variety of other types. Some distributions are bimodal (Fig. 2.6) and others are J-shaped or U-shaped. Some are symmetrical, or approximately so,

Frequency distributions 11

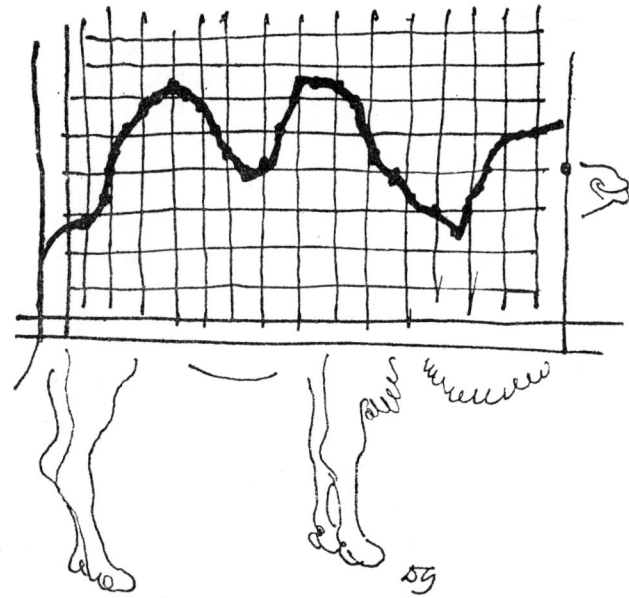

Fig. 2.6. Some distributions are bimodal
(Reproduced from *The Economist* by courtesy of the Editor)

as in Example 2.2; others are *skew*, as for instance that of Example 2.1. Often, the greater frequencies occur for the smaller values of the variable, and the skewness is then said to be positive. Example 2.1 shows negative skewness.

Representative measures

To describe a unimodal distribution by means of a few representative measures the most important features to be considered are:

(i) the position of the middle of the distribution;
(ii) the extent to which the values are spread away from the middle.

It is also possible to measure skewness, but this is less important.

For (i) the mode is not usually a convenient measure. For a discrete distribution its value may be somewhat fortuitous, and for a continuous distribution we have only a modal group, which may depend too much on the method of grouping. There are two commonly used alternatives, the *median* and the better known *arithmetic mean* or *average*.

Medians and quartiles

The *median* is the value of the variable for the middle observation when all the observations are arranged in order of magnitude. Thus, if the 629 MPs were lined up in order of age, the age of the middle one, the 315th, would be the median. (If the total frequency is an even number, the median is taken as mid-way between the two middle ones.) In Example 2.1 the 50th and 51st observations (in order of magnitude) were both of 9 references, so the median is 9. In Example 2.2, the median is the 315th (counting from either end). If we count from the lower end, there are 304 members up to and including age 50, so the median is somewhere between 51 and 56. Without knowing the exact ages of the 112 members in this group we can estimate the median value only by supposing that they are evenly distributed from 51 to 56. On this basis the median is approximately $51 + \frac{11}{112} \times 5$, or 51·5.

If the variable is X and the total frequency n, the median is the value of X for the $\frac{1}{2}(n + 1)$th observation in order of magnitude (it being understood that if $\frac{1}{2}(n + 1)$ is not an integer, the average of the two nearest values is taken). In a similar way we can define values of the variable that divide the series of observations, arranged in order, into quarters. The values for the one-quarter and three-quarter marks are called respectively the *lower quartile* and *upper quartile*. More precisely, they are the $\frac{1}{4}(n + 1)$th and $\frac{3}{4}(n + 1)$th values in order of magnitude. The difference between them affords a rough measure of the spread of the observations. It is usual to take half the difference and to describe it as the *semi-interquartile range*, or *quartile deviation*.

In Example 2.1, the lower quartile is the value for No. $25\frac{1}{4}$, counting from the lower end, i.e. 7. (The fraction $\frac{1}{4}$ presents no problem here, but in any case these are rough measures and the value for No. 25 could well be taken.) The upper quartile is the value for No. $75\frac{3}{4}$ (or say No. 76), i.e. 10. The semi-interquartile range is $1\frac{1}{2}$.

In Example 2.2, the lower quartile is the value for No. $157\frac{1}{2}$, counting from the lower end. As the first three groups contain 114 members, the lower quartile is approximately

$$41 + \frac{43\frac{1}{2}}{96} \times 5, \quad \text{or} \quad 43\cdot 3.$$

The upper quartile, reckoning from the top end, is

$$61 - \frac{32\frac{1}{2}}{88} \times 5, \quad \text{or} \quad 59\cdot 2.$$

The semi-interquartile range is, therefore, approximately 7·95.

Cumulative frequencies

In the above way of measuring the middle of the distribution and the spread we are concentrating attention on the proportion of observations for which the values are above or below a certain figure. Thus, the upper quartile is the value above which one-quarter of the readings lie. The idea can obviously be extended. We could, for example, divide the readings into 10 or 100 parts and call the corresponding values of the variable *deciles* or *percentiles*. This suggests drawing a graph to show how many readings are above or below any value.† Such a graph is called a *cumulative frequency curve*. In Example 2.2, the number of members above or below certain ages is shown in the following table, and in Fig. 2.7a, b.

Age	Frequency	Cumulative frequencies			
81+	1	Above 81	1	Below 81	628
71–81	14	71	15	71	614
66–71	25	66	40	66	589
61–66	85	61	125	61	504
56–61	88	56	213	56	416
51–56	112	51	325	51	304
46–51	94	46	419	46	210
41–46	96	41	515	41	114
36–41	66	36	581	36	48
31–36	43	31	624	31	5
26–31	5	26	629	26	0

Such treatment is appropriate when it is of interest to know what proportion of the readings are above or below a given value. For example, in considering the results of an examination, it is often required to know how many candidates, or what proportion of the candidates, have marks above a given level. The choice between the two ways of drawing the curve naturally depends on whether the interest is on the proportion above a given level or the proportion below.

A cumulative frequency curve provides a convenient means of determining quickly the median and quartiles, or the values above which any given percentage of the readings lie.

The mean

The median and the quartile deviation do not lend themselves readily to algebraical development. The arithmetic mean (usually called simply the *mean*), with its associated measure of spread, known as

† For a discrete variable it is usual to take the number down to and including (or up to and including) the given value.

14 Statistics: the how and the why

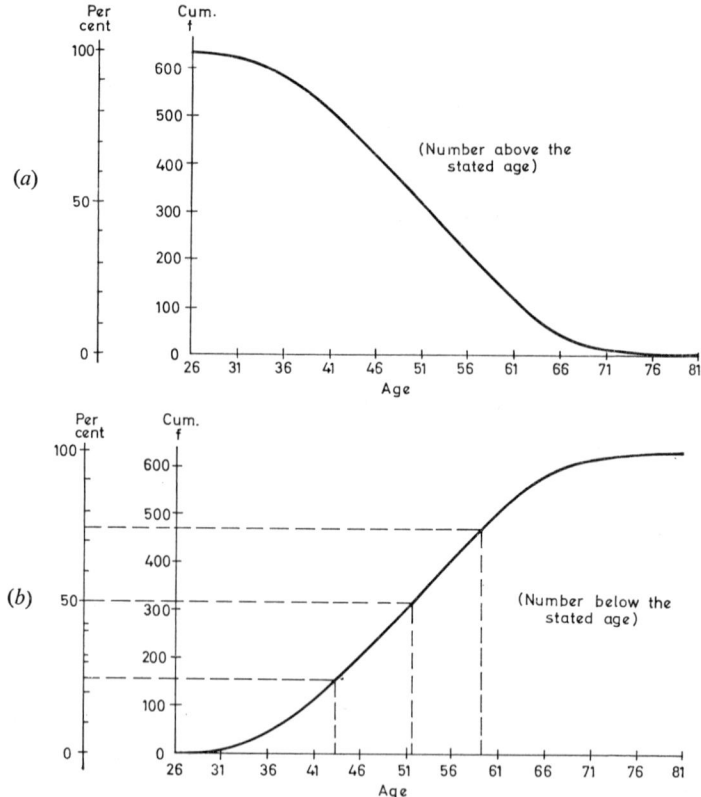

Fig. 2.7. Ages of MPs, 1964 (Cumulative frequency)

the *standard deviation*, is used much more, both in the theory of statistics and in practical applications.

If the observed values of a variable X are $X_1, X_2, X_3, \ldots, X_n$, where n is the total number of observations, the mean is denoted by m or $\bar{X}$, and is defined by

$$\bar{X} = \frac{X_1 + X_2 + X_3 + \cdots + X_n}{n} = \frac{\Sigma X}{n}.$$

The values $X_1, X_2, \ldots$ are not usually all different, and, in fact, a particular value X may occur with frequency f. It is thus sometimes convenient to write

$$\bar{X} = \frac{\Sigma fX}{n},$$

where it is understood that the summation extends over the *different* values of X. (The apparent contradiction between these two expressions for $\bar{X}$ is sometimes found confusing by beginners. This only illustrates that in statistical work, as in all mathematics, it is necessary to be quite clear about the meanings of the letters used.)

The reader who is familiar with mechanics will note the analogy with the centre of mass of a number of collinear particles, which is, of course, the mean position of the mass of the particles. As with centre of mass, the position of the mean is independent of the origin chosen and of the scale. This may be proved as follows:

If $X = a + kx$, where a and k are constant,

$$\bar{X} = \frac{\sum(a + kx)}{n} = \frac{na + k\sum x}{n} = a + k\bar{x},$$

where $\bar{x}$ is the new coordinate of the mean. In particular, if the mean is taken as the origin for x, $\bar{X} = a$ and $\sum x = 0$, i.e. the sum of the deviations from the mean is zero; or, with the mechanics analogy in mind, we may say that the sum of the moments about the mean is zero.

In a skew distribution, the median usually comes between the mean and the mode, often at about one-third of the distance from the mean to the mode. (A useful mnemonic is that, for negative skewness, the three values come in alphabetical order, with the mean and median near together, as in the dictionary.)

In calculating a mean, much arithmetic can be saved by changing the origin and scale. Thus, in Example 2.2, using X for the central value of a group and x for the value on the new scale:

Age	X	f	x	fx
26–31	28·5	5	−5	−25
31–36	33·5	43	−4	−172
36–41	38·5	66	−3	−198
41–46	43·5	96	−2	−192
46–51	48·5	94	−1	−94
51–56	53·5	112	0	−681
56–61	58·5	88	1	88
61–66	63·5	85	2	170
66–71	68·5	25	3	75
71–81	76	14	$4\frac{1}{2}$	63
81+	(83·5)	1	6	6
		629		402
				−681
				−279

$$\bar{x} = \frac{-279}{629} = -0.444.$$

But
$$X = 53.5 + 5x;$$
$$\therefore \text{ mean age} = \bar{X}$$
$$= 53.5 - 5 \times 0.444,$$
$$= 51.3 \text{ approx.}$$

The new origin is often referred to as the *assumed mean*. This is because of the obvious advantage of taking it somewhere near the true mean, to which it is thus a first approximation. The value of x for any group can be described as the 'code-number' of the group.

Variance and standard deviation

We now need a measure of dispersion. If we consider the deviations from the mean, their sum is zero (as proved above). We can, of course, suppress the minuses and find the mean of the deviations regardless of sign. This gives the *mean deviation*. But it is an unsatisfactory procedure, leading to awkward algebra. It is more convenient to square the deviations and take the mean of the squares. This gives the *variance*, defined by

$$\text{Variance} = \frac{(X_1 - \bar{X})^2 + (X_2 - \bar{X})^2 + \cdots + (X_n - \bar{X})^2}{n},$$

$$= \frac{\Sigma (X - \bar{X})^2}{n}.$$

As before, this may be calculated as $[\Sigma f(X - \bar{X})^2]/n$, the summation being taken over the *different* values of x.

The calculation may be carried out with any origin and any scale, for if $X = a + kx$,

$$\text{Variance} = \frac{\Sigma (X - \bar{X})^2}{n},$$

$$= \frac{\Sigma (a + kx - a - k\bar{x})^2}{n},$$

$$= k^2 \cdot \frac{\Sigma (x - \bar{x})^2}{n}.$$

The multiplier k^2 represents the necessary change back to the original scale.

The variance is a quantity of the second degree in x, and is denoted by s^2. Then s, the square root of the variance, is of the same dimensions as x and is called the *standard deviation* (S.D.). To avoid confusion between the scales, a suffix may be used. Thus,

$$s_X = ks_x,$$

where s_X is the standard deviation on the original scale, and s_x that on the working scale or 'code-numbers'.

The analogy with mechanics will again be noted, the standard deviation s corresponding exactly to the radius of gyration about an axis through the centre of mass.

In calculating the variance it is again an advantage to use an arbitrary origin. If x is so measured,

$$s_x^2 = \frac{\sum (x - \bar{x})^2}{n}$$

$$= \frac{\sum x^2 - \sum 2x\bar{x} + \sum \bar{x}^2}{n}$$

$$= \frac{\sum x^2 - 2\bar{x} \sum x + n\bar{x}^2}{n}$$

$$= \frac{\sum x^2 - 2\bar{x}n\bar{x} + n\bar{x}^2}{n}$$

$$= \frac{\sum x^2}{n} - \bar{x}^2.$$

$\sum x^2/n$ is sometimes called the 'second moment of the distribution about the origin'. The variance is the second moment about the mean. The formula

$$\text{variance} = \frac{\sum x^2}{n} - \bar{x}^2$$

shows that the variance is less than the second moment about any other point. Thus, the mean has the important property of being the point about which the second moment is least.

As before, the variance formula may be written

$$s_x^2 = \frac{\sum fx^2}{n} - \bar{x}^2,$$

where f is the frequency, the summation being taken over the *different* values of x.

The calculation, in Example 2.2, proceeds as follows:

X	f	x	fx	fx²
28·5	5	−5	−25	125
33·5	43	−4	−172	688
38·5	66	−3	−198	594
43·5	96	−2	−192	384
48·5	94	−1	−94	94
53·5	112	0	−681	0
58·5	88	1	88	88
63·5	85	2	170	340
68·5	25	3	75	225
76	14	4½	63	283·5
(83·5)	1	6	6	36
			402	2857·5
			−681	
			−279	

Divide by 629:   −0·444 .   4·543

$$s^2 = \frac{\Sigma fx^2}{n} - \bar{x}^2 = 4\cdot543 - (0\cdot444)^2,$$

$$= 4\cdot346.$$

Therefore $s_x = 2\cdot084$, and s_X (the standard deviation on the original scale) is 10·4 years approx.

The mean and standard deviation (now called m and s on the original scale) may be shown on the histogram by vertical lines, as in Fig. 2.8. It will be noticed that the range from $m - 2s$ to $m + 2s$ covers nearly the whole of the diagram. In this case the area outside this range corresponds to a frequency of 18, or just under 3 per cent. As a general rule it will be found that for unimodal distributions the range $m - 2s$ to $m + 2s$ covers something like 95 or 96 per cent of the total frequency (it is often called 'the 95 per cent range'); and that the wider range $m - 3s$ to $m + 3s$ covers virtually the whole distribution. It must be emphasized that this is a rough rule, but in practice it will be found to apply to a large number of distributions, even when they are moderately skew or irregular. Rough as it may be, the rule is of considerable value. In the first place it gives a good idea of the meaning of standard deviation; secondly, it provides a useful rough check on the result of the calculation.

To check the calculation accurately, the best method is to do it again with a different origin. If the two results disagree, it is useful to note that the totals of the fx columns should differ by an integral multiple of n, namely, $n(\bar{x}_1 - \bar{x}_2)$, where $\bar{x}_1, \bar{x}_2$ are the coordinates of the mean; and the totals of the fx^2 columns should differ by $n(\bar{x}_1^2 - \bar{x}_2^2)$, which is an integral multiple of $(n\bar{x}_1 + n\bar{x}_2)$, the sum of the two fx columns.

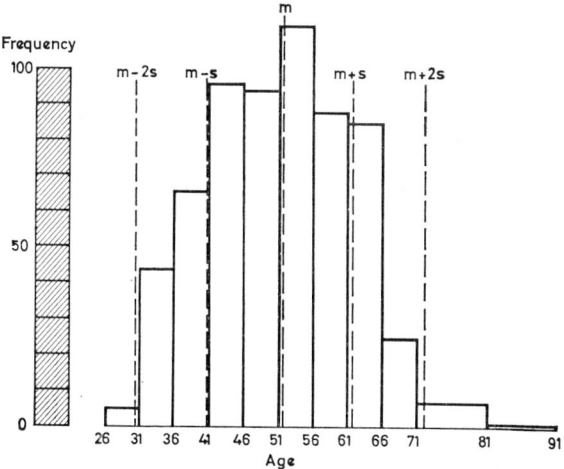

Fig. 2.8. Age distributions of MPs, 1964

The rough rule mentioned above gives an idea of the way in which standard deviation measures dispersion; that is to say, if we know the mean and the standard deviation we can specify limits between which about 95 per cent of the observations will probably lie. But the question still remains: 'What is standard deviation for?' The answer is that it gives us a standard for judging whether any particular observation is exceptional. In the case illustrated, it would probably be agreed that any member of the House older than 72 is unusually old, and any member below the age of 30·4 is unusually young. In the same way, if we take the average height of a man as 5 ft 9 in, and the standard deviation as $2\frac{1}{2}$ in, any man over 6 ft 2 in may be regarded as tall, and any over 6 ft $4\frac{1}{2}$ in as very tall.

EXERCISE 1

1. Name, and illustrate roughly, the type of diagram that could suitably be used to illustrate each of the following:
    (i) the number of gold, silver, and bronze medals won by different countries in the Olympic Games,

(ii) the relative importance (as measured by number of events) of each of the main divisions of the games (athletics, swimming, winter events, etc.),
(iii) the number of athletes (men and women) competing from the different countries,
(iv) the improvement in the time for two or three given races (e.g. the 1500 metres and the 3000 metres) since the year 1900,
(v) the ages of competitors in a particular year.

2. The monthly turnover of a small shop is as follows:

Oct.	Nov.	Dec.	Jan.	Feb.
£184	£178	£212	£170	£158

Do the figures for the last three months indicate a decline in the business?

3. The following table refers to the City and County of Bristol:

Year	Area (acres)	Population
1920	11,705	328,945
1921	18,436	376,975

Find by slide rule, desk calculator, or rough estimate the density of population in each year, and comment on the result.

4. Before a certain match, two bowlers have each taken 21 wickets for 315 runs. During the match, one of them takes 5 for 120 and the other 1 for 48. Find their new averages, and comment.

5. A class of 30 pupils have an hour each week in which they read books of their own choice. Would you expect the average length of all the books they read in a year to be the same as, or greater than, or less than, the average length of the 30 books they are reading on one particular day?

6. In 1696, Gregory King estimated that out of a total of about 5,500,000 people in England, some 600,000 were over the age of 60. Would it be right to deduce that at that time about one-tenth of the children born lived to at least 60?

7. In a physics experiment with a Geiger counter, a record was kept of the number of clicks heard per half-minute. The frequency of the different numbers recorded was as follows:

No. of clicks	2	3	4	5	6	7	8	9	10	11
Frequency	1	0	0	4	3	6	4	11	10	8
No. of clicks	12	13	14	15	16	17	18	19	20	
Frequency	12	8	3	3	3	1	1	1	2	

Find the modal value, the median, the quartiles, and the semi-interquartile range.

8. The weights of 100 boys were distributed as follows:

Weight in lb	No. of boys
100–110	4
110–120	9

Weight in lb	No. of boys
120–130	15
130–140	12
140–150	18
150–160	21
160–170	14
170–180	7

Draw a cumulative frequency diagram showing the number of boys above any given weight. Estimate the median weight, the quartiles, and the semi-interquartile range.

9. The percentage marks of 600 candidates in an examination were distributed as follows:

Marks	No. of candidates	Marks	No. of candidates
15–19	4	50–54	76
20–24	16	55–59	52
25–29	64	60–64	30
30–34	56	65–69	8
35–39	80	70–74	14
40–44	92	75–79	4
45–49	100	80–84	4

Draw a cumulative frequency curve showing the number of candidates who gained a given number of marks or more. Estimate
   (i) the median and quartiles,
   (ii) the number of candidates who got at least $62\frac{1}{2}$ per cent,
   (iii) the mark that 30 per cent of the candidates failed to reach.

10. The pulse-rates of 28 boys were measured before and after doing a certain exercise. The ratios obtained were as follows, each correct to 1 decimal place:

   1·3 1·6 1·4 1·8 1·4 1·2 2·0 1·4 1·9 1·3
   1·5 1·4 1·4 1·3 1·5 2·3 1·8 2·1 1·2 1·3
   1·9 1·2 1·4 1·1 2·2 1·1 2·1 1·4

Make a frequency table and draw a diagram. Find the mean ratio and the standard deviation, and show them on the diagram.

11. The lengths of 209 broad bean seeds (Seville Long Pod) were measured to the nearest mm, with the following results:

Length (cm)	Frequency	Length (cm)	Frequency
2·1	1	2·8	35
2·2	3	2·9	29
2·3	5	3·0	13
2·4	15	3·1	9
2·5	28	3·2	—
2·6	32	3·3	—
2·7	37	3·4	2

Find the mean length and the standard deviation.

12. Measurements were made of the width of stomatal apertures in a lilac leaf, with the following results. The measurements were in scale divisions, to the nearest $\frac{1}{4}$, one division representing 2·8 microns.

Width (scale divisions)	No. of stomata	Width	No. of stomata
1	1		
$1\frac{1}{4}$	0	$2\frac{1}{2}$	6
$1\frac{1}{2}$	5	$2\frac{3}{4}$	2
$1\frac{3}{4}$	8	3	7
2	10	$3\frac{1}{4}$	1
$2\frac{1}{4}$	9	$3\frac{1}{2}$	1

Find the mean width and the standard deviation, in microns.

13. The marks of 250 candidates in an examination were distributed as follows:

Marks	No. of candidates	Marks	No. of candidates
20–24	5	50–54	32
25–29	24	55–59	24
30–34	20	60–64	11
35–39	38	65–69	4
40–44	44	70–74	6
45–49	40	75–79	2

Draw a histogram to illustrate this distribution. Find the average mark and the standard deviation, and show these on the diagram. Test the rough rule that about 4 to 5 per cent of the candidates should have marks differing from the average by more than twice the S.D.

14. The lengths of 90 sentences from Trevelyan's *Social History of England* are shown in the following tables:

No. of words	Frequency	No. of words	Frequency
5– 9	9	40–44	3
10–14	11	45–49	1
15–19	10	50–54	1
20–24	20	55–59	2
25–29	10	60–64	3
30–34	12	65–69	—
35–39	6	70–74	1
		75–79	1

Find, approximately, the mean number of words per sentence and the standard deviation.

Make a similar count for any other author, and compare the results.

Frequency distributions 23

15. The following table shows the age distribution of men in certain P.O. Telephone construction gangs in Scotland:

Age	Percentage of men	Age	Percentage of men
15–20	1·3	40–	7·8
20–	8·8	45–	9·2
25–	19·0	50–	7·4
30–	22·6	55–	2·3
35–	19·4	60–65	2·2

Draw a diagram to illustrate the distribution. Estimate the mean age and the standard deviation, and show them on the diagram.

16. The heights of 300 lorry drivers were found to be as follows:

Mid-interval height (in)	No. of drivers	Mid-interval height (in)	No. of drivers
$63\frac{1}{2}$	8	$71\frac{1}{2}$	28
$64\frac{1}{2}$	9	$72\frac{1}{2}$	16
$65\frac{1}{2}$	18	$73\frac{1}{2}$	9
$66\frac{1}{2}$	29	$74\frac{1}{2}$	7
$67\frac{1}{2}$	35	$75\frac{1}{2}$	3
$68\frac{1}{2}$	49	$76\frac{1}{2}$	1
$69\frac{1}{2}$	53	$77\frac{1}{2}$	1
$70\frac{1}{2}$	34		

Find the mean height and the standard deviation.

17. In a biological experiment on a large number of cells the following results were obtained:

Concentration of cell sap	Percentage of such cells
Less than 0·55	0
Less than 0·60	5
Less than 0·65	20
Less than 0·675	40
Less than 0·70	80
Less than 0·75	100

Estimate the mean concentration of cell sap. (See solutions for a suggested method.)

18. The scores of 100 teams in a rifle-shooting competition were distributed as follows:

Score	No. of teams	Score	No. of teams
60– 69	2	120–129	6
70– 79	2	130–139	13
80– 89	5	140–149	15
90– 99	8	150–159	18
100–109	6	160–169	9
110–119	7	170–179	7
		180–189	2

24   Statistics: the how and the why

Draw a cumulative frequency curve and use it to find the semi-interquartile range. Use the rough rule

$$\text{standard deviation} \simeq \tfrac{3}{2} \times \text{semi-interquartile range}$$

to estimate the standard deviation. Find the S.D. by the ordinary method and compare with the estimate.

19. The monthly rainfall in England and Wales (average for the area) during 100 consecutive months is shown in the following table, the values being recorded to the nearest tenth of an inch:

Rainfall (in)	Frequency	Rainfall (in)	Frequency
0·5 and below	3	3·6–4·0	12
0·6–1·0	2	4·1–4·5	13
1·1–1·5	14	4·6–5·0	5
1·6–2·0	6	5·1–5·5	3
2·1–2·5	10	5·6–6·0	—
2·6–3·0	13	6·1–6·5	2
3·1–3·5	16	6·6–7·0	1

Find the mean monthly rainfall and the standard deviation. Show on a histogram the range $m \pm 2s$ and estimate the area of the diagram outside this range, expressing it as a percentage of the whole area.

20. Define the mean m and the standard deviation s of a variable x. Prove that $s^2 = \sum (x^2/n) - m^2$, where n is the total frequency. Prove that $\sum (x - k)^2$ is least when k is the mean.

21. If two classes, of 24 and 30 pupils respectively, have the same average age of 15 years 6 months, but standard deviations of 6 months and 8 months respectively, find the standard deviation for the combined group of 54.

22. In a certain year the average rainfall for a group of 20 stations in the northern part of a country was 775 mm, with a S.D. of 35 mm. For 20 stations in the southern part the average was 725 mm, with a S.D. of 30 mm. Find the S.D. for the combined group of 40 stations.

23. The mean score of the first 25 scripts marked by an examiner was 50 marks, with a standard deviation of $10\sqrt{2}$ marks. The next two scripts scored 80 marks each. Using 50 as an 'assumed mean', find the new standard deviation for the 27 scripts.

24. If n_1 observations give a mean of m_1 and a S.D. of s_1, and a further group of n_2 observations give a mean of m_2 and a S.D. of s_2, find the mean for the combined group of $(n_1 + n_2)$ observations and show that the variance is

$$\frac{n_1 s_1^2 + n_2 s_2^2}{n_1 + n_2} + \frac{n_1 n_2 (m_1 - m_2)^2}{(n_1 + n_2)^2}.$$

25. At a village fête there is a boy-and-girl relay race in which a boy runs 100 yards and his partner then runs back to the starting-point. There

Frequency distributions

are 10 boys and 12 girls present, and their times for 100 yards are shown in the following table:

Boys		Girls	
Time in sec.	Frequency	Time in sec.	Frequency
11	1	11	—
12	2	12	1
13	4	13	2
14	2	14	5
15	—	15	2
16	1	16	—
		17	2

Tabulate the 120 possible times for the relay race. Find the mean and the variance **(i)** for the boys, **(ii)** for the girls, **(iii)** for the relay race. Verify that the answers to **(iii)** for the mean and variance respectively are the sums of those for **(i)** and **(ii)**.

3
Lines of closest fit. Correlation

The method of least squares

It has been seen (p. 17) that the mean is the value such that the sum of the squares of the deviations from it is a minimum. This idea of measuring the combined effect of a number of deviations by the sum of their squares can be usefully extended. When the results of a scientific experiment are plotted on a graph, it frequently happens that the points lie nearly on a straight line, or on some other well-defined curve, and we may want to find the line of closest fit. For a straight line this is often done by stretching a piece of cotton among the points and finding the best position by eye, but the *method of least squares* enables us to obtain a more definite result.

The following table, and the graph (Fig. 3.1) show the results of an experiment to determine the absorption of oxygen by the lungs as the work done by the body increases:

X (work done)	300	420	680	840	954 kg m
Y (oxygen absorbed)	680	820	1000	1080	1300 cm^3/min

It is desired to draw a straight line as near as possible to the plotted points. It might be thought that we should try to minimize the sum of the squares of the perpendicular distances, but as the scales on the two axes are different there is no meaning to perpendicular distance; it would vary according to the particular scales chosen. So we must measure parallel to one or other of the axes: in other words, we must consider Y as a function of X (as here), or X as a function of Y, and compare the observed values of Y (or X) with those given by a linear equation such as $Y = mX + c$ (or $X = m'Y + c'$).

For any point (X_r, Y_r) the deviation from the value given by $Y = mX + c$ is $(Y_r - mX_r - c)$, and the sum of the squares of such

Lines of closest fit. Correlation

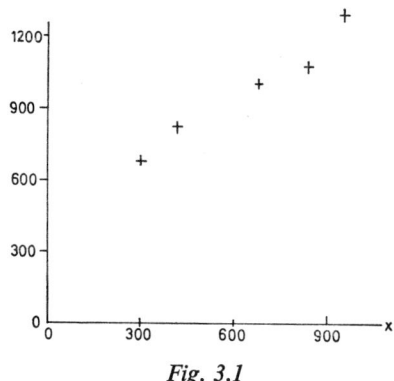

Fig. 3.1

deviations may be written $\sum (Y_r - mX_r - c)^2$. We have to choose values of m and c to make this a minimum.

Differentiating partially with respect to c and to m,

$$\sum -2(Y_r - mX_r - c) = 0, \qquad (1)$$
and
$$\sum -2X_r(Y_r - mX_r - c) = 0. \qquad (2)$$

From the first of these equations,

$$\sum Y_r - m \sum X_r - nc = 0,$$

where n is the total number of points. Dividing by n,

$$\bar{Y} - m\bar{X} - c = 0, \qquad (3)$$

where $(\bar{X}, \bar{Y})$ is the mean centre of the points. The desired line must therefore pass through the mean centre.

In view of this, it is natural and convenient to use the mean centre as origin. If the points are now (x_r, y_r), the above work still applies, but $\bar{x}$ and $\bar{y}$ are both zero. From equation (3), $c = 0$, and equation (2) becomes

$$\sum -2x_r(y_r - mx_r) = 0.$$

Hence
$$m = \frac{\sum x_r y_r}{\sum x_r^2} = \frac{p}{s_x^2},$$

where
$$s_x^2 = \frac{\sum x^2}{n}, \text{ the variance of } x,$$

and
$$p = \frac{\sum xy}{n}, \text{ called the } \textit{covariance}.$$

The line of closest fit, when we regard y as a function of x, is then

$$y = \frac{p_{xy}}{s_x^2} x, \quad \text{or} \quad Y - \bar{Y} = \frac{p_{XY}}{s_X^2}(X - \bar{X}).$$

(There has been no change of scale, so $s_x = s_X$ and $p_{xy} = p_{XY}$.) This line is called a *line of regression* (see p. 29 for the origin of the term), and its gradient p/s_x^2 is called a *coefficient of regression*.

Calculation of a regression coefficient

We already know that, in calculating s_x^2, an arbitrary origin can be used. Thus, if X is measured from the arbitrary origin and x from the mean centre,

$$s_x^2 = \frac{\sum(X - \bar{X})^2}{n} = \frac{\sum X^2 - 2\bar{X}\sum X + n\bar{X}^2}{n}$$

$$= \frac{\sum X^2}{n} - 2\bar{X}^2 + \bar{X}^2$$

$$= \frac{\sum X^2}{n} - \bar{X}^2.$$

In the same way,

$$p = \frac{\sum(X - \bar{X})(Y - \bar{Y})}{n}$$

$$= \frac{\sum XY}{n} - \bar{X}\frac{\sum Y}{n} - \bar{Y}\frac{\sum X}{n} + \frac{n\bar{X}\bar{Y}}{n}$$

$$= \frac{\sum XY}{n} - \bar{X}\bar{Y}.$$

For the figures given above, we can use the original origin, with a simple change of scale. (But s_x^2 and p must, of course, be measured on the original scales.)

X	Y	X'	Y'	X'²	X'Y'
300	680	30	68	900	2 040
420	820	42	82	1 764	3 444
680	1 000	68	100	4 624	6 800
840	1 080	84	108	7 056	9 072
954	1 300	95·4	130	9 101	12 402
Totals		319·4	488	23 445	33 758
Divide by 5:		63·9	97·6	4 698	6 751·6

$$s_X^2 = 100(4689 - 63\cdot9^2) \qquad p = 100(6752 - 63\cdot9 \times 97\cdot6)$$
$$= 100(4689 - 4083) \qquad\qquad = 100(6752 - 6237)$$
$$= 60\,600. \qquad\qquad\qquad = 51\,500.$$
$$\text{Gradient} = p_{XY}/s_X^2 = 515/606 = 0\cdot850.$$

The line of closest fit is therefore
$$Y - 976 = 0\cdot850(X - 639)$$
or
$$Y = 0\cdot850X + 433.$$

Bivariate distributions

A bivariate distribution is one in which, for each member of the 'population', there are values of two variables to be considered. We could, for example, classify the 629 MPs by weight as well as by age, and it might be of interest to enquire whether there would be any association between the two. (This is not the same situation as was considered above, where the values of one variable could be chosen arbitrarily. In the experiment on absorption of oxygen, the work done could be arranged by the experimenter, only the oxygen absorbed being found by observation. The 629 MPs were chosen by their membership of the House and the values of both variables would be determined by observation.)

A bivariate distribution may be illustrated by a *scatter diagram*. Figure 3.2a represents the marks of 80 candidates in two mathematics papers, the coordinates of each dot being the marks of one particular candidate in Paper I and Paper II. The dots cover a roughly elliptical area, owing evidently to the tendency of the better candidates to do well in both papers, and of the weaker ones to do badly in both. If there were no such association between the marks on the two papers, one would expect the area covered to be more nearly symmetrical about lines drawn parallel to the axes.

The two lines l_1 and l_2 are lines of closest fit, drawn as described above, l_1 for y considered as a function of x, l_2 for x as a function of y. If we consider the groups of dots in successive columns, we see that the mean positions of the groups follow roughly the line l_1. Similarly, if we follow the successive rows, the mean value of x for each row is near to the line l_2.

It will be noticed that, as x increases, the mean value of y (represented by l_1) increases, but not so fast as x does; and similarly, as y increases, the mean value of x increases, but not so fast as y does. For this reason the lines l_1 and l_2 are called *lines of regression*, l_1 being the line of regression of y on x, and l_2 the line of regression of x on y. The term originates from Sir Francis Galton's study of the inheritance of

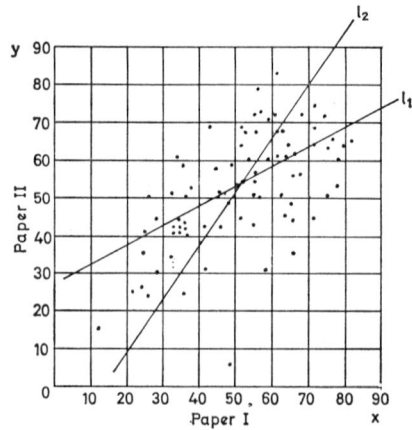

Fig. 3.2a. Marks in two mathematics papers

stature (*Natural Inheritance*, 1889) in which he found that the children of tall parents were, on the average, less tall than their parents, which suggested a regression towards the norm; but this does not mean that the population is gradually approaching a uniform height, because it is equally true that the parents of tall children are, on the average, less tall than their children. This paradox is explained by the scatter diagram, the two effects being represented by the gradients of the lines l_1 and l_2.

If there were a more perfect association (or *correlation*) between x and y, the dots all lying for example near to a straight line, the regression lines would nearly coincide. If there were no association, they would be at right angles and parallel to the axes; and if there were a tendency for y to decrease as x increased (called *negative correlation*), the lines would be even wider apart, and in the extreme case would again coincide.

Correlation

The gradients of the regression lines provide a means of measuring correlation. If the lines coincide, the gradients (measured with respect to the x and y axes respectively) are reciprocals. Then

$$\frac{p}{s_x^2} \times \frac{p}{s_y^2} = 1,$$

and
$$\frac{p}{s_x s_y} = +1 \text{ or } -1,$$

Lines of closest fit. Correlation

according as the correlation is positive or negative. This suggests that a *coefficient of correlation* r might be defined as $p/(s_x s_y)$.

It should be noted that, under this definition, r will not depend on the scales on which the variables are measured, since the numerator and denominator of the fraction are of the same degree in x and in y. Moreover r will be zero only when $p = 0$, as would be true for a distribution symmetrical about lines parallel to the axes.

It may further be proved that r lies between -1 and $+1$. For, if m is the gradient of l_1,

$$\sum (y_1 - mx_1)^2 = \sum y_1^2 - 2m \sum x_1 y_1 + m^2 \sum x_1^2,$$

$$= ns_y^2 - 2 \cdot \frac{p}{s_x^2} \cdot np + \frac{p^2}{s_x^4} \cdot ns_x^2,$$

$$= ns_y^2 - n\frac{p^2}{s_x^2},$$

$$= ns_y^2(1 - r^2).$$

As $\sum (y_1 - mx_1)^2$ is necessarily positive, r^2 is less than 1, and r lies between -1 and $+1$.

Thus r is in every way suitable as a measure of correlation. It is independent of the scales chosen, and ranges in value between $+1$ and -1, these values corresponding to perfect positive and negative correlation.

Calculation of r

The practical calculation of a correlation coefficient may be somewhat laborious, but can be shortened by grouping and by the use of arbitrary scales. Thus, in the example illustrated in Fig. 3.2 we make a table (Fig. 3.2b) arranged in the same manner as the scatter diagram:

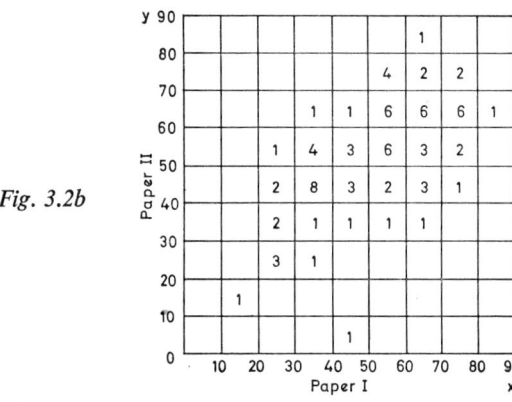

Fig. 3.2b

32   Statistics: the how and the why

fy²	fy	f	y									fxy		
9	3	1	3						1			6		= 6
32	16	8	2					4	2	2		8 + 8 + 12		= 28
21	21	21	1			1	1	6	6	6	1	−1 + 6 + 12 + 18 + 4		= 39
−	−	19	0		1	4	3	6	3	2		−		
19	−19	19	−1		2	8	3	2	3	1		4 + 8 − 2 − 6 − 3		= 1
24	−12	6	−2		2	1	1	1	1			8 + 2 − 2 − 4		= 4
36	−12	4	−3		3	1						18 + 3		= 21
16	−4	1	−4	1								12		= 12
25	−5	1	−5			1						−		
182	−12	80		−4	−3	−2	−1	0	1	2	3	4	x	111
			80	−	1	8	15	9	19	16	11	1	f	
			54	−	−3	−16	−15	−	19	32	33	4	fx	
			254	−	9	32	15	−	19	64	99	16	fx²	

Fig. 3.2c

The boundary values, strictly speaking, are $9\frac{1}{2}$, $19\frac{1}{2}$, etc., but there is practical convenience in labelling them 10, 20, etc. It will not affect the result. We now proceed, in fact, to re-label the groups in an arbitrary manner and to calculate s_x^2, s_y^2, and p_{xy} in terms of the new scales (x and y) (Fig. 3.2c).

$$\bar{x} = \frac{54}{80} = 0.675 \qquad s_x^2 = \frac{254}{80} - (0.675)^2 = 2.72$$

$$\bar{y} = \frac{-12}{80} = -0.15 \qquad s_y^2 = \frac{182}{80} - (0.15)^2 = 2.25$$

$$p_{xy} = \frac{111}{80} - (-0.15)(0.675) = 1.50$$

$$\therefore r = \frac{1.50}{\sqrt{(2.72 \times 2.25)}} = 0.61.$$

The significance of different values of r is a difficult question. The most that can be said at the present stage is that values above 0.5 are quite high, provided they are based on a large number of observations (say over 50).†

† An interpretation of r may be given as follows: If $y_i = mx_i + e$ (i.e. if e is the

Lines of closest fit. Correlation

When the number of readings is comparatively small, it may be easier to arrange the work in a table similar to that on p. 28, perhaps grouping the figures by the simple device of omitting the units digit of each, as in the next example.

Example 3.1
Twenty-one candidates were examined in Latin and English, their marks being as shown.

Latin	English	x	y	x^2	xy	y^2
108	48	10	4	100	40	16
93	46	9	4	81	36	16
66	48	6	4	36	24	16
30	29	3	2	9	6	4
35	26	3	2	9	6	4
87	43	8	4	64	32	16
55	31	5	3	25	15	9
24	31	2	3	4	6	9
54	28	5	2	25	10	4
128	65	12	6	144	72	36
65	38	6	3	36	18	9
128	64	12	6	144	72	36
100	28	10	2	100	20	4
168	67	16	6	256	96	36
112	44	11	4	121	44	16
162	50	16	5	256	80	25
142	71	14	7	196	98	49
131	51	13	5	169	65	25
138	34	13	3	169	39	9
34	33	3	3	9	9	9
140	43	14	4	196	56	16
		191	82	2149	844	364
Divide by 21:		9·10	3·90	102·3	40·19	17·33

divergence of a point from the regression line, measured in the y-direction), it follows from the equation on p. 31 that

$$\frac{1}{n}\sum e^2 = \frac{1}{n}\sum (y_i - mx_i)^2 = s_y^2(1 - r^2),$$

i.e.   variance of $e = s_y^2(1 - r^2)$.

So r^2 represents the proportion of the y-variance that is removed by measuring from the regression line instead of from $\bar{y}$.

Hence $s_x^2 = 102\cdot3 - 9\cdot10^2 = 19\cdot5,$
$s_y^2 = 17\cdot33 - 3\cdot90^2 = 2\cdot12,$
$p_{xy} = 40\cdot19 - 9\cdot10 \times 3\cdot90 = 4\cdot70,$

and $r = \dfrac{4\cdot70}{\sqrt{(19\cdot5 \times 2\cdot12)}} = 0\cdot73$ approx.

(This is a high value of r and likely to be significant, even though the number of readings is small.)

It should be noted that a significant value of r does not necessarily imply any causality between the variables. Thus it might be that the number of sailing dinghies in the country, taken over the past ten years, would show a high correlation with the number of computers, but no one would suppose that either influenced the other.

Sometimes two variables not directly related are both influenced by the same factor, and a high value results. For example, increases in the numbers of motor cars and washing machines might both be due to growing affluence.

EXERCISE 2

1. A bar is supported at its ends in a horizontal position. When a weight of x kg is suspended from the mid-point of the bar the greatest deflection of the bar is y cm. Measured values of y for various values of x are as follows:

x	20	30	40	50	60	70	80
y	0·51	0·90	0·98	1·31	1·35	1·60	1·91

Find **(i)** the means of x and y, **(ii)** the variance of x, **(iii)** the covariance of x and y, **(iv)** the equation of the line of regression of y on x. Plot a graph and draw the regression line.

2. The weight of a side of bacon when cured is related to the weight of the live pig before it is killed. The following table gives figures for Berkshire pigs:

Live weight of pig	124	152	187	201	219	lb
Weight of cured side of bacon	30	40	50	60	70	lb

Find a regression equation giving the weight of a side of bacon in terms of that of the live pig.

3. The following table shows the output of the different coalfields in 1956 and the number of men employed:

	Output (million tons) y	No. of industrial workers (thousands) x
Scottish	22	90
Northern	14	51
Durham	26	111
North-eastern	44	143
North-western	16	63
East Midlands	46	108
West Midlands	18	59
South-western	24	111
South-eastern	2	7

Find an equation for the regression of y on x and illustrate by a graph.

4. The mark y obtained out of a possible 70 by each of 200 children is paired with the child's age x months. The results are shown in the following grouped frequency table:

Mark (y)	Age in months (x)						
	120	123	126	129	132	135	138
0–9	1	5	3				
10–19		4	11	2			
20–29		5	13	16	3	2	
30–39		2	4	21	13	7	3
40–49		1	3	12	16	8	4
50–59				4	6	11	3
60–69				3	2	4	8

For each central value of x calculate the mean mark obtained. Plot these mean marks against x on a graph, and sketch in a regression line of y on x. (Ignore the first point.) Obtain the equation of the regression line in the form $y = mx + c$.

If it were desired to adjust the marks obtained so as to make allowance for age, how many marks should be added to those obtained by a child aged 126 months and how many should be subtracted from those obtained by a child aged 134 months? (Take 129 months as the standard age.)

5. The average price, for each of eight successive months, of a certain commodity, in dollars in New York (y) and in pounds sterling in London (x), is given in the following table:

y	83	85	88	80	98	95	97	94
x	28	29	29	27	31	32	32	32

Find (i) the covariance of x and y, (ii) the equation of the line of regression of y on x, (iii) the correlation coefficient.

6. The following table gives the number of road vehicles x (unit 1,000,000) in use in September of each year and the number of road casualties y (unit 100,000) in each year:

Year	Vehicles (x)	Casualties (y)
1952	4·9	2·1
1953	5·3	2·3
1954	5·8	2·4
1955	6·4	2·7
1956	6·9	2·7
1957	7·5	2·7
1958	7·9	3·0
1959	8·6	3·3

Find the correlation coefficient between x and y.

7. The following table gives indices of employment and business activity for the years 1929–37, the year 1935 being taken as standard in each case:

	1929	1930	1931	1932	1933	1934	1935	1936	1937
Business activity (x)	98·5	93	87·5	84	89	96	100	106	112
Employment (y)	98·5	94·5	91	90	93·5	97·5	100	105	111

Tabulate the deviations from 100 and find an equation for the regression of employment on business activity.

8. Flour from various strains of wheat was milled and baked under uniform controlled conditions, the object being to determine whether there was any correlation between the percentage of protein in the various flours and the size of the loaf when baked. The results were as shown below:

Loaf volume in cm³	Protein per cent	Loaf volume in cm³	Protein per cent
1980	14·9	1990	13·4
2030	14·3	2010	14·1
2235	17·1	2060	14·1
2245	15·4	2000	14·1
2285	15·3	1930	14·9
2225	14·7	1970	13·7
2030	14·3	1980	13·7
2070	14·7	2010	13·6
2010	14·1	2020	13·4
2000	13·9	2010	13·9

Make a scatter diagram and divide it into squares whose sides represent 50 cm³ and 0·5 per cent. Proceed as in Fig. 3.2c and find the correlation coefficient.

9. The following table shows the annual rainfall and sunshine recorded at Felsted from 1912 to 1937:

Year	Rainfall (in)	Sunshine (hr)	Year	Rainfall (in)	Sunshine (hr)
1912	26·37	1459	1925	24·83	1602
1913	18·86	1454	1926	24·03	1374
1914	24·71	1693	1927	29·21	1480
1915	25·49	1675	1928	25·33	1665
1916	31·26	1468	1929	20·51	1712
1917	23·67	1692	1930	23·80	1517
1918	25·76	1659	1931	23·32	1329
1919	24·71	—	1932	20·12	1371
1920	24·00	1663	1933	15·23	1605
1921	14·41	1931	1934	19·40	1468
1922	21·30	1606	1935	26·55	1544
1923	22·74	1519	1936	23·19	1155
1924	31·99	1618	1937	27·75	1314

Neglecting the year 1919, find the correlation coefficient between the rainfall and sunshine. (Take groups covering 1 inch and 100 hours.)

4
Probability

Our everyday estimates of probability are based on one or other of three different methods. First, there is the purely subjective judgement that certain events are equally likely, or that one event is more likely to occur than another.† We may, perhaps, go further and say that one is twice as likely as the other. These judgements are often expressed in terms of 'odds', as in such phrases as 'a fifty-fifty chance' or 'ten to one it won't happen'. They are generally coupled with a firm, if vague, belief that the 'equally likely' events will, in the long run, occur equally often.

Somewhat similar is the second method, in which, again, a number of events are judged to be equally likely, and the question is then asked how many of them lead to a particular result. Thus, if a pack of cards is dealt out to four players, it is judged that the ace of spades is equally likely to go to any of the four and that the probability of a particular player having it is, therefore, 1 in 4 (corresponding to odds of 3 to 1 against).

The third method is based on relative frequencies. If, at a certain place, rain is recorded on 750 days out of 1000, we commonly consider that there is a three-quarters chance of rain occurring on any one particular day. This is not entirely remote from the second method, since if one of the 1000 days is picked at random there is, according to the second method, a probability of 750 in 1000 that rain will have fallen on that day. Where this method differs from the second is in the supposition that this probability, based on past experience, will hold good for future days.

To build a theory of probability on frequencies, it is necessary to make an assumption that, as the number of observations is increased, the relative frequency of a particular result approaches a limit. This is an assumption: it cannot be proved. But that it is not entirely wide

† J. M. Keynes built an entirely non-numerical theory of probability concerned with the logical relationships of such judgements.

of the mark will appear later (p. 75). If the assumption is granted, the limit approached can be taken as the measure of probability, and the observed relative frequency in a large number of trials can be taken as an approximation to it.

The theory of probability put forward in this book is based on the second method, which allows an easier development and provides an adequate basis for statistical theory. If, on any occasion, a probability is estimated from observed frequencies, it will be in the knowledge that a new assumption is being made.

Definition of probability

If an event can happen in n ways and fail in n' ways, all these ways being equally likely, then the probability (or chance) of its happening is $n/(n + n')$.

An immediate consequence of the definition is that probabilities range from 0 to 1, 0 representing impossibility and 1 certainty. The probability of the event's failing is, by definition, $n'/(n + n')$, and the sum of the probabilities of its happening and failing is 1.

If $P(A)$ is the probability that an event A will happen, and $P(\overline{A})$ is the probability that it will not happen,

$$P(A) + P(\overline{A}) = 1.$$

Two events. Conditional probability

If two events A and B are considered, there are four possibilities: (i) that both should happen, (ii) that A should happen and B should

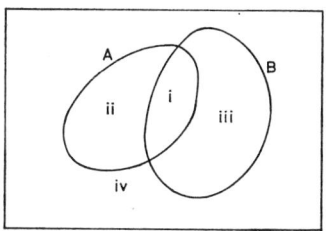

Fig. 4.1

not, (iii) that B should happen and A should not, (iv) that neither should happen. The corresponding probabilities may be written as

(i) $P(A \cap B)$, (ii) $P(A \cap \overline{B})$, (iii) $P(\overline{A} \cap B)$, (iv) $P(\overline{A} \cap \overline{B})$.

The Venn diagram in Fig. 4.1 represents the ways in which the events can happen or not happen. It can also be taken as representing the corresponding probabilities.

If we wish to know the probability that A and B should both happen, we can first consider the ways in which A can happen (represented by areas (i) and (ii) in the diagram), and then consider what proportion of these ways is included in area (i) (Fig. 4.1).

Suppose there are n ways in which A can happen, and n' ways in which it can fail, all equally likely. Suppose further that, of the n ways in which A can happen, r involve the happening of B. Then

$$P(A \cap B) = \frac{r}{n + n'}$$

$$= \frac{r}{n} \times \frac{n}{n + n'}$$

$$= \frac{r}{n} \times P(A).$$

r/n is called the *conditional probability* of B happening, given that A happens; or, more briefly, 'of ~~A given B~~ B given A', and is denoted by $P(B \mid A)$. Thus

$$P(A \cap B) = P(B \mid A) \times P(A),$$

and, similarly,
$$= P(A \mid B) \times P(B). \qquad (1)$$

Example 4.1

If one of the numbers 1, 2, 3, ..., 40 *is picked at random, the chance that it is divisible by* 3 *is* $\frac{13}{40}$. *Of the 13 numbers divisible by 3 only two (namely 15 and 30) are divisible by 5. So the conditional probability that one of these numbers known to be divisible by 3 should also be divisible by 5 is* $\frac{2}{13}$, *and the probability of the number being divisible by both 3 and 5 is* $\frac{13}{40} \times \frac{2}{13}$, *i.e.* $\frac{2}{40}$, *as can also be seen more directly.*

The probability that A OR B will happen is denoted by $P(A \cup B)$. It can be seen from the diagram that this is equal to

$$P(A) + P(B) - P(A \cap B). \qquad (2)$$

Alternatively, it is
$$1 - P(\bar{A} \cap \bar{B}).$$

In Example 4.1, the probability that the number chosen should be divisible by 3 or 5 is $\frac{13}{40} + \frac{8}{40} - \frac{2}{40}$, i.e. $\frac{19}{40}$, as may be verified by direct counting.

Mutually exclusive events

If A and B cannot both happen together, i.e. if $P(A \cap B) = 0$, they are said to be *mutually exclusive*. The identity (2) then becomes

$$P(A \cup B) = P(A) + P(B).$$

For example, the probabilities that the number chosen from the first 40 should be divisible by 7 and 6 are $\frac{5}{40}$ and $\frac{6}{40}$ respectively. But there is no number less than 42 that is divisible both by 7 and by 6. So the probability of the number chosen being divisible by 7 or 6 is $\frac{5}{40} + \frac{6}{40}$, i.e. $\frac{11}{40}$.

Independent events

The events A and B are said to be independent if the happening of either does not affect the probability of the other, i.e. if

$$P(A \mid B) = P(A) \quad \text{and} \quad P(B \mid A) = P(B).$$

The equations (1) then become

$$P(A \cap B) = P(A) \times P(B). \qquad (3)$$

It is evident, in the light of equations (1), that any one of these three statements implies the other two.

It also follows, in such a case, that

$$P(A \mid B) = P(A \mid \bar{B}),$$
for
$$P(A \cap \bar{B}) = P(A) - P(A \cap B),$$
$$= P(A) - P(A).P(B),$$
$$= P(A)\{1 - P(B)\},$$
$$= P(A).P(\bar{B}).$$

Hence, $\qquad P(A \mid \bar{B}) = P(A) = P(A \mid B).$

In other words, the probability of A's happening is the same whether B is known to have happened, or known not to have happened, or not known about at all.

Example 4.2

A die is thrown and, at the same time, a card is drawn from a pack of ten consisting of 3 aces, 3 kings, 3 queens, and a knave. Points are scored for an ace, or a six, or for both together. There are 60 possible combinations, distributed as follows:

		Cards		
		Ace	Not-ace	Totals
Die	Six	3	7	10
	Not-six	15	35	50
	Totals	18	42	60

The probability of drawing an ace is $\frac{18}{60}$, which is the same as either of the conditional probabilities $\frac{3}{10}$ and $\frac{15}{50}$. The probability of throwing a six is, similarly, $\frac{1}{6}$. The probability of drawing an ace *and* throwing a six is $\frac{3}{60}$, which is the same as $\frac{3}{10} \times \frac{1}{6}$.

Independence, as defined above, is purely a matter of the number of cases that can occur; the term *statistical independence* is sometimes used to make this clear. The events A and B need not be entirely disconnected. Thus, if a die is thrown, the possibilities of the resulting number being divisible by 2 or by 3 are events that might seem to be not entirely independent, in the everyday sense of the word; but they are, in fact, statistically independent, since if the number is divisible by 3 there is one possibility out of two that it is even, and if it is not divisible by 3 there are two possibilities out of four. The probability of its being even is, thus, independent of whether or not it is a multiple of 3.

Three or more events

The above rules for addition and multiplication of probabilities can be extended to three or more events by successive application. (For multiplication, the second event must be independent of the first, the third must be independent of the various possible outcomes of the first two, and so on.)

If three events A, B, C are not mutually exclusive, the appropriate Venn diagram shows that

$$P(A \cup B \cup C) = P(A) + P(B) + P(C) \\ - P(B \cap C) - P(C \cap A) - P(A \cap B) + P(A \cap B \cap C).$$

Readers familiar with the algebra of sets will be able to give a formal proof of this, and to extend it by induction. In general,

$$P(A \cup B \cup C \cup \cdots) \\ = \sum P(A) - \sum P(A \cap B) + \sum P(A \cap B \cap C) - \cdots$$

Permutations and combinations

It will be evident that the difficulty of more complicated examples in probability will be largely that of counting the number of ways in which events can turn out. As the reader probably has some knowledge of permutations and combinations, a brief summary only is given here.

The number of *permutations* of n things r at a time (denoted by $_nP_r$) is the number of ways of choosing r things in order from n things available (no thing being chosen twice).

The first thing can be chosen in n ways, and for each of these ways the second can be chosen in $(n-1)$ ways, the third in $(n-2)$ ways, and so on. Thus,

$$_nP_r = n(n-1)(n-2)\ldots(n-r+1),$$
$$= n!/(n-r)!,$$

where, $\quad n! = 1.2.3.\ldots,n. \quad$ and $\quad 0! = 1.$

In particular, the number of ways in which n things can be arranged among themselves, $_nP_n$, is $n!$. If repetitions are allowed, the number of permutations of n things r at a time is n^r.

The number of *combinations* of n things r at a time (denoted by $_nC_r$) is the number of selections of r things, out of n things available, no regard being had to the order in which the r things are picked. It is evident that a combination of r things can be arranged in $r!$ orders. Therefore, each combination makes $r!$ permutations, and hence

$$_nC_r = {}_nP_r \div r!,$$
$$= \frac{n(n-1)(n-2)\ldots(n-r+1)}{1.2.3.\ldots r},$$
$$= \frac{n!}{r!(n-r)!}.$$

Both the last two forms are useful. The second exhibits a symmetry as between r and $(n-r)$, which corresponds to the fact that in making a selection of r things we automatically make one of $(n-r)$. Thus, $_nC_r = {}_nC_{n-r}$, and, in general, each is equal to the number of ways of dividing n things into two groups containing respectively r things and $(n-r)$.

The exception to the last statement is when the two groups are equal in numbers. They are then interchangeable, and each mode of division corresponds to two combinations. For example, there are six ways of picking two people out of four, namely AB, AC, AD, BC, BD, CD. But there are only three ways of dividing the four people into two pairs, namely AB and CD, AC and BD, AD and BC.

The number of ways of dividing n things into groups of $p, q, \ldots,$ where $p + q + \cdots = n$, is, in general, $\dfrac{n!}{p!q!\ldots}$.

Proof: The first group can be chosen in $_nC_p$ ways, the second in $_{n-p}C_q$ ways, and so on. Hence, the total number of ways is

$$\frac{n!}{p!(n-p)!} \times \frac{(n-p)!}{q!(n-p-q)!} \times \cdots = \frac{n!}{p!q!\ldots}.$$

Exceptions again occur if some or all of the groups contain equal numbers of things. If m of the groups contain the same number, each possible division will have been counted $m!$ times, and, hence, the result must be divided by $m!$. Thus, for example, the number of ways of dividing six things into three groups of two is $\dfrac{6!}{2!2!2!3!}$.

Application to probability

Generally speaking, a problem in probability can be solved by either of two methods: by 'counting of ways', using the formulae of permutations and combinations, or by combining probabilities according to the rules for addition and multiplication described above. It is often advisable to check one method by means of the other.

Example 4.3

If two cards are drawn from an ordinary pack of 52, what are the chances (a) that both will be aces, (b) that one will be an ace and the other not, (c) that neither will be aces?

By the first method. The number of possible pairs is $_{52}C_2$, and the number of possible pairs of aces is $_4C_2$. So the probability of two aces is $_4C_2/_{52}C_2$, or $\dfrac{4.3}{52.51}$.

The number of possible pairs containing one ace is 4×48; so the second answer is $4 \times 48/_{52}C_2$, or $\dfrac{8.48}{52.51}$.

Similarly, the third answer is $_{48}C_2/_{52}C_2$, or $\dfrac{48.47}{52.51}$.

By the second method. The chance of drawing an ace the first time is $\tfrac{4}{52}$. There are then 51 cards left, including 3 aces. So the chance of drawing an ace the second time is $\tfrac{3}{51}$. As these are independent events, the combined chance is $\dfrac{4.3}{52.51}$.

Probability 45

For (b) there are two possibilities: an ace followed by a non-ace, for which the chance is $\frac{4}{52} \times \frac{48}{51}$, or a non-ace followed by an ace, for which it is $\frac{48}{52} \times \frac{4}{51}$. These are mutually exclusive possibilities, so we add the chances, obtaining the same answer as before. For (c) the answer, by a similar argument to that used for (a), is $\frac{48}{52} \times \frac{47}{51}$.

It will be noticed that, in this example, it makes no difference whether we think of the two cards as being drawn simultaneously (as in the first method) or successively (as in the second method). But it would be a different problem if the first card were replaced before a second one was drawn. Then, the chances for two aces, one ace, or none would be

$$\frac{4}{52} \times \frac{4}{52}, \quad 2 \times \frac{4}{52} \times \frac{48}{52}, \quad \frac{48}{52} \times \frac{48}{52}.$$

These are the terms of the expansion of $\left(\frac{4}{52} + \frac{48}{52}\right)^2$.

Successive trials. The binomial distribution

When successive trials are made, and the chances of success or failure are the same in each, the chances of the different possible numbers of successes are given by the terms of a binomial expansion. If p is the chance of success and q the chance of failure at each of n trials, the chances of $0, 1, 2, \ldots, n$ successes are given by the terms in the expansion of $(q + p)^n$, i.e. for

0,   1,   2, r, n successes

the chances are

$$q^n, \quad nq^{n-1}p, \quad {}_nC_2 q^{n-2}p^2, \ldots, \quad {}_nC_r q^{n-r}p^r, \ldots, \quad p^n.$$

Proof: The chances of success in r particular trials and of failure in the remaining $(n - r)$ trials (these being all independent of each other) is $p^r q^{n-r}$. But the r particular trials might be chosen in ${}_nC_r$ ways. Therefore, the total probability of r successes and $(n - r)$ failures is ${}_nC_r q^{n-r}p^r$.

It will be noticed that the sum of all these chances is $(q + p)^n$, i.e. 1, since $q + p = 1$.

An alternative way of describing the binomial distribution is to say that the probability of obtaining exactly r successes is the coefficient of t^r in $(q + pt)^n$, or in $(1 - p + pt)^n$. This function of t is sometimes called a *probability generating function*. The letter t is, of course, a dummy, used merely as a convenient way of labelling the different probabilities.

EXERCISE 3

Permutations and combinations

1. (i) How many ways are there of forecasting 'win', 'lose', or 'draw' for 10 football matches?
   (ii) In how many ways can first, second, and third prizes be awarded among 10 competitors?
   (iii) In how many ways can 3 oranges be distributed among 10 children if no orange is cut and no child is to have more than one?

2. In how many ways can 32 oranges be distributed among 18 children, if no orange is cut, every child is to have at least one, and no child is to have more than two?

3. In how many ways can 15 oranges be distributed among 10 children if no orange is cut, and no child is to receive more than two?

4. In how many ways can a set of tennis be made up from 6 boys and 4 girls if a boy and a girl play on each side? In how many of these ways will one particular boy (i) find himself playing, (ii) find himself playing against a particular girl?

5. If n switches can each be put in any of three positions A, B, or C, in how many ways can p of them be in position A, q in position B, and r in position C (where $p + q + r = n$)?
   If there are 9 such switches, in how many ways can at least 6 of them be put in position A?

6. In how many ways can 22 people be divided into two cricket teams (i) to play as 1st XI and 2nd XI against another club, (ii) to play against each other in a 'friendly' game?

7. A committee of 8 decides to send a deputation to see the Home Secretary. In how many ways can the deputation be chosen from among the 8 members.

8. Ten articles are to be placed in a row, three of them, A, B, and C, coming together. Prove that this can be done in 241,920 ways.
   In how many ways can 9 articles be arranged in a row so that two of them, A and B, do not come together?

9. Ten coloured beads are to be arranged on a table in a circle. In how many ways can this be done (i) when all the beads are of different colours, (ii) when three are of the same colour and the rest different?

10. A family of 12, consisting of 6 boys and 6 girls, meet together for Christmas.

Probability

On the first day of Christmas, each gives each a present. How many presents are given?

On the second day of Christmas, they play six-a-side hockey. In how many ways can they divide into two teams?

On the third day, four are invited out to dinner. In how many ways can the four be chosen?

On the fourth day they go to church, 5 in a car, 4 on bicycles and 3 on foot. In how many ways can they be divided for this?

On the fifth day they dance. In how many ways can the six boys choose their partners for one dance?

On the sixth day they play chess, six games taking place simultaneously. In how many ways can they be divided for this?

On the seventh day some of the party leave and some stay on ('some' meaning 'one or more'). In how many ways can this happen?

11. Show that the number of combinations, $(n - 2)$ at a time, of n things of which three are alike and the rest different (n being greater than 3) is $\frac{1}{2}(n^2 - 5n + 8)$.

12. (i) A school of 6 houses plays cricket leagues, each house playing against each other. How many matches will there be? How many ways are there of arranging (a) the first round, (b) the second round?
    (ii) Answer the same questions when there are 8 houses.
    (iii) A house has 15 players, and the rules are that no one shall miss two matches running and no one shall play more than twice running if other players who have not played twice running are available. How many possible teams are there for the first match, for the second, the third, and the fourth?

13. (i) In how many ways can 6 people be seated at a 'round' table? (The term 'round' must be taken to mean that there is no distinction between one seat and another.)
    (ii) If there are 6 seats at a 'round' table, in how many ways can 3 persons seat themselves?
    (iii) In how many ways can 6 different keys be arranged on a ring?
    (iv) In how many ways can 6 keys, of which 3 are alike and the rest different, be arranged on a ring?
    (v) A molecule of benzene (C_6H_6) can be represented by a hexagon, and substitution products can be represented by placing appropriate letters at the vertices of the hexagon; thus, a di-substitution product can be formed in 3 ways. In how many ways can a tri-substitution product be formed, using three different atoms?

14. If $ABCDEF$ is a hexagon inscribed in an ellipse, we know by Pascal's Theorem that the three intersections of opposite sides, namely the intersections of AB and DE, BC and EF, CD and FA, lie on a straight line. Show that the six points could be joined to form 60 different hexagons and that their Pascal lines will intersect, 4 at a time, in 45 points.

48   Statistics: the how and the why

15. If an octagon is inscribed in a conic and the sides are numbered 1, 2, 3, 4, 5, 6, 7, 8 in order, the eight intersections of sides 1 and 4, 2 and 5, 3 and 6, etc., lie on another conic. Show that, given eight points on a conic, they can be joined to form 2520 different octagons and that these will determine 2520 other conics intersecting, 96 at a time, in 210 points.

Probability

16. If two dice are thrown, what is the probability of (i) a double six, (ii) a total score of 8?

17. In a certain district, 50 per cent of the voters support party A, 30 per cent party B, and 20 per cent party C. If an enquirer chooses three voters at random, find as decimals the chances:

    (i) that all three should be supporters of A,
    (ii) that two should be supporters of A and one of B,
    (iii) that there should be one of each party.

    If 5 voters are picked at random, what is the chance that they will not all be of the same party?

18. A man has 3 florins and 5 pennies in his pocket. If he takes out three coins at random, what is the chance of all three being pennies? If, instead, he takes out three in succession, replacing each before drawing the next, what is then the chance of all three being pennies?

19. At a children's party, presents are given to 4 of the boys and 5 of the girls, chosen in each case by drawing names out of a hat. If there are 8 boys and 12 girls, what is the chance that a boy and his twin sister, who are both at the party, should both receive presents? What is the chance that one of them, but not both, should receive a present?

20. Three archers shoot at a target, their probabilities of success being $\frac{1}{4}$, $\frac{1}{3}$, and $\frac{1}{5}$. Find the probability (i) that exactly two of them will hit the target, (ii) that at least two will do so.

21. An electric circuit contains three fuses which have independent probabilities p, p, and p' of blowing when the circuit is switched on. Prove that the probability that exactly one fuse blows when the circuit is switched on is $(1 - p)(2p + p' - 3pp')$. Determine the probability that exactly two fuses blow.

22. From a squad of 20 soldiers, four are to be selected for a fatigue, one of the four to be in charge of the party.

    (i) In how many ways can the choice be made?
    (ii) If they are picked at random, what is the chance that a particular man will be in the party?
    (iii) If there are two brothers in the squad, find the probability that they will both be in the party, one of them in charge.
    (iv) If a party of four is selected from the squad on two successive days, the choice being at random on each occasion, find the

probability that each of the brothers will be picked once and only once, but not both on the same day.

23. Six white balls and four black balls, which are indistinguishable apart from colour, are placed in a bag. If six balls are taken from the bag, find the probability of their being three white and three black.
24. Two different integers are chosen at random from the first 60 integers. Find the chance that their sum is (i) even, (ii) a multiple of 3.
25. Three boys and three girls arrive separately for a party. Find the chance that no two girls arrive in succession.
26. A and B throw 3 dice alternately. A throws first, and his total score is 6.

   (i) Find the chance that B's first throw will not total more than 6.
   (ii) Find the chance that B's third throw should be the first to be more than 6.

27. Two boys A and B throw a die alternately, A having the first throw. The first to throw a six gets a prize, but if neither succeeds in two throws each, the prize goes to a third boy C. Find the chance that A gets the prize and the chance that C gets it.
28. The 18 competitors for a hundred yards race are divided into 3 heats of 6 runners each, the best two from each heat to compete in the final.

   (i) Find the chance that two particular runners should be in the same heat.
   (ii) Find the chance that the third best runner should fail to reach the final.
   (iii) Find the chance that the best three runners should all win their heats.

29. The competitors in a school boxing competition are drawn from six houses, two of which are at a distance from the gymnasium, so that their competitors must travel by bus. Find the probability that a bout should involve someone who must travel by bus (i) if in each weight there is only one competitor from each house, (ii) if in each weight there are n from each house, the rules being such that two members of the same house may box against each other.
30. There are six teams entered for a 'knock-out' competition. Two draw byes in the first round, the byes being placed one in each half of the draw. Assuming that each team plays true to form, find (i) the probability that the second-best team will reach the final, (ii) the probability that the third-best team will reach the semi-final.
31. An insurance policy provides for the payment of £100 at the end of one year if either of two persons, x aged 50 and y aged 60, die within the year. The probability that a life aged 50 should survive a year is 0·984, and the probability for a life aged 60 is 0·970. Find the probability that the £100 will be payable.

What would the probability have been if the £100 had been payable (i) only in the event of both persons dying within the year, (ii) only in the event of one but not both dying within the year?

(Check the three answers by considering the relation between them.)

32. Five cards are drawn from a pack, each being replaced before the next is drawn. Represent by a binomial expansion the chances that 5, 4, 3, 2, 1 or none of them should be hearts. Evaluate the chances of drawing (i) exactly three hearts, (ii) at least three hearts.

33. Four people play bridge together every day for a week, partners being determined each evening by cutting. What is the chance that Mr A should draw Mrs B as a partner on exactly 4 of the 7 occasions? What is the chance that it should happen on more than 4 of the 7 occasions?

34. In a game of Piquet there are 32 cards, of which 12 are dealt to each hand. Of the remaining 8 cards, the holder of the 'elder hand' may then draw five. Assuming that he does so, prove that the odds are 3 to 1 against his drawing any particular card not already in his hand. If the holder of the 'younger hand' then takes the remaining three cards, find the corresponding odds in his case.

35. If two cards are drawn from an ordinary pack of 52, find the chance that one should be an ace and the other a ten, knave, queen, or king. Show that the odds against it happening twice running are more than 400 to 1.

36. A bookmaker offers to take bets at the following odds for a race in which six horses are running:

Delphi 5 to 1
Earnalot 3 to 1
Upjenkins 5 to 1
Royal Banner 4 to 1
Mount Vernon 1 to 1
Van Dieman 8 to 1.

Show that the sum of the corresponding chances is greater than 1, and suggest a reason for this. The bookmaker tries to arrange that the total of his possible gains and losses on each horse is the same. Show that if he succeeds in doing this he will make the same profit whichever horse wins.

37. Twelve people are chosen at random for an opinion quiz. Assuming that they are equally likely to be men or women, find the probability that at least 9 should be men. If, in fact, there are 9 men and 3 women, find the probability that exactly 3 of the men and 2 of the women should have been born on a Sunday.

38. Two tennis players, A and B, play against each other every day, A winning, on an average, two days out of three. Find the probability that on seven particular days A should win at least five times.

Probability

39. A botanist collects 15 plants of a certain species, and of these 15 plants 11 display a feature the others lack. He wishes to consider whether this is consistent with the hypothesis that two-thirds of such plants display the feature and one-third do not. If the chance of any plant displaying the feature were two-thirds, what would be the chance of obtaining 10 with it, and 5 without it, in a sample of 15? What would be the chance of obtaining 11 with it, and 4 without it? Do you think that the observation of 11 with the feature is evidence against the hypothesis of a two-thirds to one-third proportion?

40. A jury is chosen from a 'population' of possible jurymen who would, if empanelled, be equally divided as to their verdict (guilty or not guilty) in a particular case. Find the chances
  (i) that a jury of 12 would convict unanimously,
  (ii) that a jury of 9 would do so,
  (iii) that a jury of 12 would convict by a majority of at least 10 to 2,
  (iv) that a jury of 12 would convict by a majority of at least 9 to 3.

41. Show that the chance of a bridge hand being a 'Yarborough', i.e. consisting entirely of cards of denomination 2, 3, 4, 5, 6, 7, 8, 9, is approximately 1 in 1800.

Expectation

In practical affairs we are guided not only by our estimates of the probability of various events but also by the advantages or disadvantages that will consequently accrue to us. Thus a motorist, in deciding whether to use a road that is liable to be flooded, will balance the probable saving of time if he gets through successfully against the small chance of a considerable loss of time if the road is flooded. A gambler will willingly pay a moderate sum in exchange for a small chance of a large profit; and, similarly, an insurer will pay a small premium to cover a remote chance of a heavy loss. The product of an amount that may be gained (whether of money or anything else) and the probability that such a thing will happen is called the *expectation*. When there is a number of different possible gains, each with its own probability, the expectation is defined as the sum of all such products. If on throwing a die a player is to receive 10s. for a six, 5s. for a five, and nothing otherwise, his expectation is $\frac{1}{6} \times 10s. + \frac{1}{6} \times 5s.$, i.e. 2s. 6d. Whether it is worth his while to pay 2s. 6d. for the privilege is, of course, another matter, depending on such considerations as whether he enjoys gambling and whether he is likely to continue gambling until a considerable loss causes him to stop. If he is asked to pay 3s. for the same game he can reasonably conclude that someone is hoping to make a profit.

When there is a probability of a loss, the expectation is negative.

Thus, if a player is to receive a shilling if a coin falls 'heads' and to pay 6d. if it falls 'tails', his expectation is $\frac{1}{2} \times 1s. - \frac{1}{2} \times 6d.$, i.e. 3d.

*One of the oldest types of problem in probability (see p. 2) is the division of stakes between two players in an unfinished game. Suppose, for example, that A and B play at dice, each throwing a pair of dice in turn, taking the pool if they throw a double, and adding one dollar to the pool if they do not. When they stop play there are 10 dollars in the pool and it would be A's turn next. How should the money be divided?

On A's first throw his expectation is $(\frac{1}{6} \times 10) - (\frac{5}{6} \times 1)$ dollars; on his second throw, $(\frac{5}{6} \times \frac{5}{6} \times \frac{1}{6} \times 12) - (\frac{5}{6} \times \frac{5}{6} \times \frac{5}{6} \times 1)$; on his third throw, $(\frac{5}{6} \times \frac{5}{6} \times \frac{5}{6} \times \frac{5}{6} \times \frac{1}{6} \times 14) - (\frac{5}{6} \times \frac{5}{6} \times \frac{5}{6} \times \frac{5}{6} \times \frac{5}{6} \times 1)$; and so on.

If S_1 is the sum of the terms on the left, and S_2 of those on the right,

$$S_1 - \frac{25}{36} S_1 = \frac{10}{6} + \frac{2 \times 25}{6^3}\left(1 + \frac{5^2}{6^2} + \frac{5^4}{6^4} + \cdots\right),$$

and

$$S_2 = \frac{5}{6}\left(1 + \frac{5^2}{6^2} + \frac{5^4}{6^4} + \cdots\right).$$

Hence

$$S_1 = \frac{960}{121}, \quad S_2 = \frac{30}{11},$$

and A's total expectation, $S_1 - S_2$, is $\frac{630}{121}$, or 5·21 dollars. A should therefore receive 5·21 dollars.

The idea of expectation is also used in calculating the value of a life annuity. Suppose that a man aged 60 is to be paid £100 if he survives one year and a further £100 if he survives a second year, interest being reckoned at 5 per cent. If his chances of surviving 1 year and 2 years are 0·9748 and 0·9494 respectively, his expectation is £97·48 at the end of one year and £94·94 at the end of the second year. The present value of his expectation is then

$$£\frac{97·48}{1·05} + \frac{94·94}{(1·05)^2},$$ which is equal to £178·95.

*Expectation of life

It is to be expected that people of a given age will continue to live some for a longer, and some for a shorter, time. The probabilities of living to different ages are unknown and can be estimated only from

Probability 53

past experience. If, out of l_{x_0} people alive at age x_0, it is reckoned that l_x will still be alive at age x, the number dying between ages x and $x + \delta x$ is $-\delta l_x$. This group will have survived approximately $x - x_0$ years. Hence, the mean number of years of survival for people aged x_0 is

$$\operatorname{Lt} \sum_{l_{x_0}}^{0} (x - x_0)(-\delta l_x) \div l_{x_0}, \quad \text{or} \quad \int_{0}^{l_{x_0}} (x - x_0) \, \mathrm{d}l_x \div l_{x_0}.$$

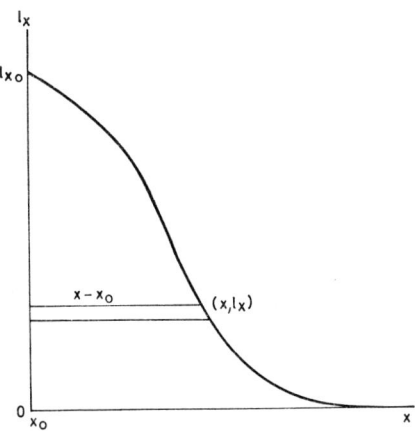

Fig. 4.2

This is represented in Fig. 4.2 by the mean 'width' of the area under the curve. It is equal to $\int_{x_0}^{\infty} l_x \, \mathrm{d}x \div l_{x_0}$, as may be seen from the figure. (The curve is a cumulative frequency curve of deaths at different ages.)

Expectation of life is seldom used in actuarial calculations, because those calculations are complicated by the compound interest element. The present value of a life annuity, for example, is not equal to that of an annuity certain for the 'expected' number of years. The reason for this is that the different possible lengths of life need to be 'weighted' according to the appropriate present values (which do not increase uniformly). A very simple example will make this clear:

Present value of £1 per annum at 5 per cent
        for 10 years   £7·7217
               15   10·3797
               20   12·4622

If three people buy life annuities and survive for 10, 15, and 20 years respectively, the mean of the survival times is 15 years, but the mean value of their annuities is less than £10·3797.

EXERCISE 4

1. A man insures a house, value £8000, and contents, value £2000, against loss by fire. If it is reckoned that during the period covered there is a chance of 1 in 1250 of a total loss, of 1 in 500 of a loss of half the contents, and of 1 in 200 of a minor claim of average amount £50, determine the appropriate premium (without allowing for expenses, profit, or interest).
2. In a 'lucky dip' there are 200 counters, one of which gives the holder a prize of £10; 25 of the others give prizes of 10s. Find the expectation if one counter is drawn.
3. If there are 24 counters in a bag, of which one gives a prize of 10s. and three others prizes of 6s., find the expectation **(i)** if one counter is drawn, **(ii)** if two are drawn together.
4. A commuter travels home by a train that gives him a two-thirds chance of catching a bus by which he can complete his journey for a fare of 1s. If he misses the bus he has to take a taxi costing 6s. Find the expected cost for any one day.
   Write down the probabilities of his catching the bus 0, 1, 2, 3 times in a period of three days. Use these probabilities to find the expected cost for the period and verify that it is three times that for a single day.
5. An annuity of £500 per annum is to be paid as long as a man aged 65, or his wife aged 60, is alive. Find the present value of the payment due at the end of the first year, given that the chance for the man to survive the year is 0·956 and for his wife 0·970, interest to be reckoned at 4 per cent per annum.
6. A coin is spun 5 times. Write down the probabilities of there being 0, 1, 2, 3, 4, 5 heads, and find the 'expected' number of heads.
7. If for each shot fired at a target there is a $\frac{2}{3}$ chance of hitting the 'bull', write down the probabilities of 0, 1, 2, 3, 4, 5, 6 successes in six shots and find the expected number of successes.
   Note. It will be proved in Chapter 6 that, if n is the number of trials and p is the probability of success in each, the expected number of successes is np.
8. Show that the probability that a six-figure number (in which any of the six digits may be 0) should contain repeated digits (i.e. two or more successive digits the same) is $1 - (\frac{9}{10})^5$; and find how many such numbers should be 'expected' out of 288 chosen at random. (Out of 288 six-figure numbers selected by the machine 'Ernie' for Premium Bond prizes, 115 were found to contain repeated digits.)

Show also that the probability of three or more consecutive digits being the same is 0·037, and find the expectation for the 288 numbers. (In the same series chosen by 'Ernie' there were 10 such numbers.)

*9. A radioactive substance decays at a rate such that the number of atoms remaining after time t is $n_0 e^{-t/k}$, where n_0 and k are constants. Show that the expectation of life of an atom is k.†

*10. If, in a community, 1000 people are born each year (uniformly spread over the year) and if the number of them still alive at age x is given by $l_x = 1000 - 10x$, show that the expectation of life is 50 years and that the population at any time consists of 50,000 people whose average age is $33\frac{1}{3}$ years.

If, with 1000 born each year, the law of survival is $l_x = \frac{1}{10}(100 - x)^2$, show that the expectation of life is $33\frac{1}{3}$ years and that the population consists of 33,333 people of average age 25.

† The expectation of life of a small bird (sparrow or robin) is said to be about 1·2 years, independent of the age of the bird. Dr H. M. Cundy (*Mathematical Gazette*, **373**, p. 294) has pointed out that this implies an analogy with radioactive decay and that the situation follows from the two premises that birds are approximately immortal but are killed off by accidents at a rate proportional to their numbers and independent of their age.

5
Probability distributions

Frequencies and probabilities

A frequency distribution represents the distribution of a 'population', i.e. a set of values of a variable. (The term 'population' is used in statistics in this technical sense. Thus, in Example 2.2 of Chapter 2, the 'population' consisted not of the members of parliament but of their ages.) If from a population one member is chosen at random (i.e. in such a way that all members are equally likely to be chosen), the probabilities of the different values of the variable are proportional to the frequencies.

If, for example, a pack of ten cards consists of 4 aces, 3 kings, 2 queens, and a knave, and one card is picked at random, the probabilities of its being an ace, a king, a queen, or a knave are $\frac{4}{10}$, $\frac{3}{10}$, $\frac{2}{10}$, and $\frac{1}{10}$ respectively.

If a histogram is drawn to represent the frequency distribution, the total area of the columns represents the total frequency. The same histogram represents the probability distribution when one member is chosen at random, the areas of the columns representing the probabilities on a scale such that the total area is 1.

Probability distributions

Suppose that a variable x can take values $x_1, x_2, \ldots$, with probabilities $p_1, p_2, \ldots$, where $p_1 + p_2 + \cdots = 1$. Such a variable is said to have a *probability distribution* and is called a *random variable* or *variate*. The 'expected value' of x, denoted by $E(x)$ or μ, is defined by

$$E(x) = \mu = \sum (px). \tag{1}$$

This corresponds to the mean of a frequency distribution, which was defined by

$$\bar{x} = \sum (fx) \div n,$$

where n was the total frequency. (The total probability is, of course, 1.)

The variance of a probability distribution is the expected value of $(x - \mu)^2$. It is denoted by σ^2 and is given by

$$\begin{aligned}
E(x - \mu)^2 = \sigma^2 &= \sum p(x - \mu)^2, \\
&= \sum (px^2) - 2\mu \sum (px) + \mu^2 \sum p, \\
&= \sum (px^2) - 2\mu^2 + \mu^2, \\
&= \sum (px^2) - \mu^2. \quad (2)
\end{aligned}$$

This corresponds to $\quad s^2 = \dfrac{\sum (fx^2)}{n} - \bar{x}^2.$

Independent variates

Two variates x and y are independent if the probability that x takes a particular value x_i is unaffected by y taking any particular value y_j, i.e.

$$P(x_i \mid y_j) = P(x_i) \quad \text{for all } i, j.$$

Then $\quad P(x_i \cap y_j) = P(x_i) \times P(y_j); \quad$ (p. 41, (3))

or, if p_i is the probability of x taking the value x_i, and q_j is the probability of y taking the value y_j, the probability that both $x = x_i$ and $y = y_j$ is $p_i q_j$.

Sum and product of independent variates

If x and y are independent variates

$$E(x + y) = E(x) + E(y),$$

and

$$E(xy) = E(x).E(y).$$

Proof: Suppose that there are n possible values for x and m for y, the corresponding probabilities being

$$p_1, p_2, \ldots, p_n \quad \text{for} \quad x_1, x_2, \ldots, x_n,$$
$$q_1, q_2, \ldots, q_m \quad \text{for} \quad y_1, y_2, \ldots, y_m.$$

There will be nm possible values for $(x + y)$ and the probability for $x = x_r$ and $y = y_s$ is $p_r q_s$.

$$\begin{aligned}
E(x + y) &= \sum_r \sum_s p_r q_s (x_r + y_s) \\
&= \sum_r \sum_s p_r q_s x_r + \sum_r \sum_s p_r q_s y_s \\
&= \sum_r p_r x_r + \sum_s q_s y_s \\
&= E(x) + E(y).
\end{aligned}$$

58 Statistics: the how and the why

The result is easily extended to any number of variates.

Again, $$E(xy) = \sum_r \sum_s p_r q_s x_r y_s$$
$$= (\sum_r p_r x_r)(\sum_s q_s y_s)$$
$$= E(x).E(y).$$

The variance of $(x + y)$ is equal to the sum of the variances of x and y.

Proof:

$$\begin{aligned}
\text{Var. } (x + y) &= E(x + y - \bar{x} - \bar{y})^2, \\
&= E(x + y)^2 - (\bar{x} + \bar{y})^2, \text{ by (2) above,} \\
&= E(x^2) - \bar{x}^2 + E(y^2) - \bar{y}^2 + E(2xy) - 2\bar{x}\bar{y}, \\
&= \text{Var. } x + \text{Var. } y + 0, \\
&\qquad\qquad \text{since } E(xy) = E(x).E(y) = \bar{x}\bar{y}, \\
&= \text{Var. } x + \text{Var. } y.
\end{aligned}$$

This result, too, is easily extended.

Difference of two independent variates

A slight modification of the above proofs shows that

$$E(x - y) = E(x) - E(y)$$

and $$\text{Var. } (x - y) = \text{Var. } x + \text{Var. } y,$$

i.e. the mean, or expected value, of the difference is equal to the *difference* of the means, and the variance of the difference is equal to the *sum* of the variances.

Example 5.1

A man who travels daily by 'underground' has to change trains each day at two stations. At the first of these he has to wait for a mean time of 3 minutes, with standard deviation 1 minute; and at the second for a mean time of 5 minutes, with standard deviation 2 minutes. Find the mean and S.D. of the total time he waits per day.

$$\text{Mean} = 3 + 5 \text{ min} = 8 \text{ min.}$$
$$\text{Var.} = 1^2 + 2^2 = 5;$$
$$\therefore \text{S.D.} = \sqrt{5} = 2\cdot24 \text{ min, approx.}$$

Members of the same population

The above example was concerned with the sum of members of two different populations; but it frequently happens that we want to pick two or more members from the same population and consider their sum or their difference or their average. The variables to be added or subtracted are not then truly independent unless it is possible for the same member to be picked more than once. If, for example, two cards are drawn from a pack, the first being replaced and the pack shuffled before the second is drawn, the two choices are independent; but if there is no replacement, it is not so. It will, however, be shown later (p. 135) that this point can be safely ignored if the population is large compared with the number of members chosen.

Suppose that $z = x_1 + x_2 + \cdots + x_n$, where $x_1, x_2, \ldots, x_n$ are n values drawn from a large population whose mean is μ and whose variance is σ^2. Then z will be distributed about a mean $n\mu$, with variance $n\sigma^2$. This follows from the theorems on pp. 57, 58, since

$$E(z) = E(x_1) + E(x_2) + \cdots,$$
$$= n\mu;$$

and similarly   Var. $z = n\sigma^2$.

Distribution of the mean of a sample

Let $\bar{x}$ be the mean of a sample of n members, so that

$$\bar{x} = (x_1 + x_2 + \cdots + x_n)/n.$$

Then each of the quantities $x_1/n, x_2/n, \ldots$, has a probability distribution with mean μ/n and variance σ^2/n^2. It follows that $\bar{x}$ is distributed about a mean μ, with variance σ^2/n. The standard deviation is therefore $\sigma/\sqrt{n}$.

The meaning of this result is that, if many such samples were taken, their means would spread away from the mean of the population much less than the population itself (Fig. 5.1). This is in line with the common sense notion that an average is more reliable than a single observation. The idea is now expressed in more definite form, and we see, for example, that to double the reliability (in the sense of making an error half as big equally likely) it is necessary to take an average of four times as many observations; or to treble it, the average of nine times as many.

In view of the importance of this result, a more compact proof is now given:

Let the mean of the population be chosen as origin, so that

$$\mu = E(x) = 0 \quad \text{and} \quad \sigma^2 = E(x^2).$$

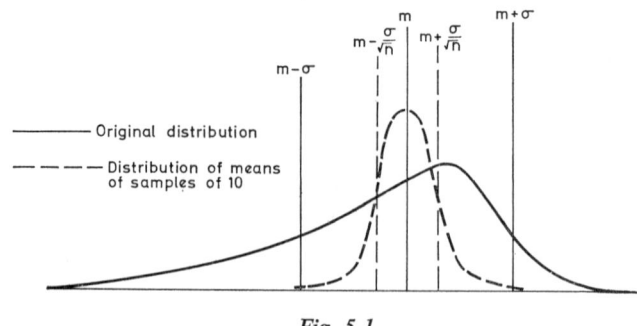

Fig. 5.1

Let $\bar{x}$ be the mean of a sample of n members, so that

$$\bar{x} = (x_1 + x_2 + \cdots + x_n)/n.$$

For the distribution of $\bar{x}$,

$$\text{the mean} = E(\bar{x}) = E(x_1/n) + E(x_2/n) + \cdots + E(x_n/n),$$
$$= \mu/n + \mu/n + \cdots + \mu/n,$$
$$= \mu = 0.$$

and

$$\text{the variance} = E(\bar{x}^2) = E\left(\frac{x_1 + x_2 + \cdots + x_n}{n}\right)^2,$$
$$= \sum_i E\left(\frac{x_1}{n}\right)^2 + \sum_i \sum_j E\left(\frac{2x_i x_j}{n^2}\right),$$
$$= n \cdot \frac{\sigma^2}{n^2}, \quad (\text{since } E(x_i x_j) = E(x_i)E(x_j) = 0)$$
$$= \frac{\sigma^2}{n}.$$

Significance tests

As already explained, the standard deviation of a distribution sets a standard by which we can measure deviations from the mean. It tells us what deviations can be regarded as exceptional or very exceptional. We have a rough rule that only about 5 per cent of a population are likely to show deviations greater than twice the standard deviation. We can now extend the same idea to the sum or average of several observations.

When a sample of n observations is taken and their average found, the result may differ from the mean of the population by an amount, great or small, which can usefully be compared with $\sigma/\sqrt{n}$, where σ is the standard deviation for the whole population. If it exceeds $2\sigma/\sqrt{n}$ it may be considered as exceptional, and if it exceeds $3\sigma/\sqrt{n}$ it is very exceptional.

The mean of a sample can be regarded as an approximation to the mean of the population, and the difference between the two is sometimes spoken of as the 'error' of the sample mean. Hence $\sigma/\sqrt{n}$, the standard error of the sample mean, is called the *standard error of the mean*. If the difference or 'error' is more than twice the 'standard error' (in absolute value), it may be regarded as 'significant', or 'significant at the 5 per cent level', and if more than three times, 'highly significant'. This rough rule will be considered more fully in later chapters.

Confidence intervals

The same idea may be expressed somewhat differently as follows:

If $\bar{x}$ is the mean of a sample of n values, the '95 per cent range' for $\bar{x}$ is given by

$$\mu - \frac{2\sigma}{\sqrt{n}} < \bar{x} < \mu + \frac{2\sigma}{\sqrt{n}},$$

i.e. $\bar{x}$ will lie within this range in about 95 per cent of all possible samples.

If we re-arrange the inequality in the form

$$\bar{x} - \frac{2\sigma}{\sqrt{n}} < \mu < \bar{x} + \frac{2\sigma}{\sqrt{n}},$$

we can say that μ lies within this range in about 95 per cent of possible cases. This range is called the *95 per cent confidence interval* for μ, meaning that, if a value of $\bar{x}$ is known, we can say with 95 per cent confidence that μ lies within the range. This will be considered further in Chapter 9.

Example 5.2

If the mean height of men in England and Wales is 5 ft 9 in, with a S.D. of 2·5 in, and a sample of 100 men from a particular county gives a mean of 5 ft 9·8 in, does the result provide evidence that men from that county are taller than the average?

The S.D. for the mean of 100 observations is $2·5/\sqrt{100}$, i.e. 0·25 in. The observed deviation is 0·8 in, which is more than three times the S.D. and thus highly significant.

The meaning of this result is worth reconsidering. Suppose there were no difference between men of that county and the rest of the population. (This supposition is called the *null hypothesis*.) Even then, the average height of a sample of 100 men might very well not be exactly 5 ft 9 in. In fact, if a large number of such samples were taken, we should expect their means to differ from 5 ft 9 in with S.D. 0·25 in. This would be due to the luck of the draw. It might conceivably happen that one sample would in this way show a deviation of 0·8 in, but as this is more than three times the S.D. (or standard error) of the mean, it would be very exceptional. We say that the difference is highly significant, meaning that we regard it as strong evidence against the null hypothesis.

If we wish to estimate the mean height of men in the county from which the sample was drawn, we must make the assumption that the standard deviation for men in that county is the same as that of the general population, namely 2·5 in. We can then say, with 95 per cent confidence, that the mean height for that county is between 5 ft 9·3 in and 5 ft 10·3 in.

Example 5.3

A man using a certain type of razor blade finds from observations over a long period that a blade lasts, on the average, 9·2 days, with a S.D. of 1·5 days. He changes to another type of blade and the first 10 blades he uses give an average of 10·0 days; the next 10 give an average of 9·8 days. What conclusion should he draw?

The null hypothesis is that there is no difference between the two types of blade. The standard deviation (or standard error) in the mean of a sample of 10 is then $1·5/\sqrt{10}$, i.e. 0·474. After the first 10 blades have been used, the mean of the sample shows a deviation of 0·8, which is rather less than twice the standard error. He must, therefore, suspend judgement. After another ten blades have been used, the deviation for the sample of 20 is 0·7; but the standard error for a sample of 20 is $1·5\sqrt{20}$, i.e. 0·337. The deviation is nearly 2·1 times the S.E., and this counts as significant. It now begins to look as if the new type of blade lasts better than the old, but a third sample would still be desirable. If it gave much the same result, say a deviation of 0·76, it would strengthen the case considerably. (The S.E. for a sample of 30 is 0·274, and the deviation for the sample, 0·72, would be nearly 2·7 times as much.)

It will be noted that these results are very much in line with commonsense conclusions, but they are more precise and more reliable.

An alternative way of treating this question is to suppose that the new blades have a different mean from the others, but the same standard deviation (a not unreasonable assumption). Then, after the first 10 have been tested, the 95 per cent confidence interval for the new mean is $10·0 \pm 2 \times 1·5/\sqrt{10}$, or $10·0 \pm 0·948$, i.e. 9·052 to 10·948. After another 10 have been tested, the interval is $9·9 \pm 2 \times 1·5/\sqrt{20}$, or $9·9 \pm 0·674$, i.e. 9·236 to 10·574.

EXERCISE 5

1. If the mean height of men is 5 ft 9 in, with a standard deviation of 2·5 in, and the mean height of women is 5 ft 4 in, with a S.D. of 2·3 in, what would be the mean total height and, hence, the mean average height of a married couple, assuming that height has no bearing on matrimonial choice? What standard deviations would the total height and the average height show?

2. The following table shows the time taken by an examiner in marking the papers of 72 candidates in statics and dynamics.

Time in min (to nearest min)	Frequencies Statics	Dynamics
1–5	1	0
6–10	5	7
11–15	11	22
16–20	17	15
21–25	13	13
26–30	10	6
31–35	7	6
36–40	3	1
41–45	3	1
46–50	1	0
51–55	1	1

   Find the mean and the standard deviation for each subject separately, and, hence, for a pair of papers, one in each subject, chosen at random.

3. A commuter drives to the station five days a week, his mean time being 20 minutes, with a S.D. of 2 minutes. For his return journey in the evening the mean time is 18 minutes, with a S.D. of 1·5 minutes. Find the variance and the standard deviation for the total time he spends on these journeys in a five-day week. State extreme limits outside which this total would hardly ever lie.

4. A group of 50 motor drivers were observed driving along a clear

section of restricted (30 mile/hr) road and their average speeds were recorded. The mean value for the 50 results was 30·0 mile/hr with a S.D. of 3·2 mile/hr. Another group of 50 gave the same mean, 30·0 mile/hr, but with S.D. 2·3 mile/hr. Find the S.D. for the combined group of 100. (Data from S. W. Quenault, *Road Research Laboratory Report* LR 70, by permission of the Director of Road Research.)

(*Note the difference between this type of question and those above.*)

5. In an examination the average mark, for all candidates taken together, is 56·3 per cent, with a S.D. of 14·0 per cent. The 75 candidates from one school have an average of 58·4 per cent. Is this significantly better than the general average?

6. A poultry keeper found that eight-week-old chickens given standard food had a mean weight of 28·3 oz, with standard deviation 2·3 oz, these figures being the result of many observations. He tried a change of diet with 48 birds and at eight weeks they had an average of 29·4 oz. Should this be regarded as reliable indication that the new diet would result in a general increase in weight? State the 95 per cent confidence limits for the mean weight of birds fed on the new diet.

7. A mass-produced component has a mean weight of 36 g, with S.D. 0·6 g. Say, with reason, whether you consider the following results show significant deviations:

(i) A random sample of three members gives a mean of 37 g.
(ii) A random sample of 25 members gives a mean of 36·2 g.

8. Steel bars made in a factory are found, by testing over a long period, to have a mean breaking strength of 33 tons per sq. in, with a S.D. of 0·36 tons per sq. in. What action would you recommend:

(i) if a batch of 50, when tested, gives an average breaking strength of 32·9 tons per sq. in,
(ii) if a further batch of 50 gives exactly the same result?

9. An automatic machine is set to cut a metal bar to lengths of 5·00 cm; it works to a standard deviation of 0·0015 cm. A sample of 36 pieces exceed 5·00 cm by the following amounts (in units of 0·0001 cm):

−6	+10	−14	−16	−10	+14	−10	+13	+10
−12	−20	−22	+21	−15	+11	−15	−10	+16
−18	+13	−14	−15	+10	+6	−10	−5	−6
−9	+20	+12	+11	+19	−15	−6	−21	−18

Calculate the mean of these deviations and compare it with the standard error of the mean. State what action, if any, you would recommend.

10. In the study of an old book it is found that the average length of the sentences is 22·6 words, with S.D. 8·2. It is suspected that a certain

Probability distributions

section of the book is an insertion by an editor. If the section contains 30 sentences and their average length is 20·3 words, does this provide supporting evidence?

11. Golf balls are manufactured to weigh 45·33 g, with a standard deviation not exceeding 0·24 g. Periodical tests are made by weighing samples of 10 balls. Between what limits would you expect the total weight of a sample to lie in 95 per cent of the tests? What action would you recommend if the weight of a sample were found to lie just outside these limits?

12. Buses run on a certain route at intervals of 15 minutes. At one stopping-point the mean number of minutes late is 1·5, with standard deviation 0·8. Determine the standard deviation of the interval between two consecutive buses. Would you regard it as exceptional if the interval between two particular buses were as short as 13 minutes?

13. A machine cuts metal rods of required length, its accuracy being such that the standard deviation is 0·06 cm. How large a sample should be examined to find the mean length of the rods almost certainly to one place of decimals?

Frequency curves

In a frequency diagram for a continuous variable, the frequency for any group is represented by the area of a rectangle. The height of the rectangle represents the *frequency-density*. For example, there were 25 members of the House of Commons between the ages of 66 and 71 (p. 10) and 14 between 71 and 81, giving frequency densities of 5 and 1·4 respectively (Fig. 2.5, p. 9). If the numbers for each year of age were known, and the corresponding diagram drawn, the heights would probably form an irregular sequence. But in cases where the total frequency is sufficiently large, and the group intervals small, the tops of the rectangles may approximate to a smooth curve. This is called a *frequency curve*. Such curves may approximate to certain standard algebraic types, and their statistics can then be evaluated theoretically.

If a frequency curve has the equation $y = f(x)$, the ordinate y represents the frequency-density and the element of area $y \, \delta x$ represents the frequency corresponding to the interval δx.

The total frequency n is given by

$$n = \int f(x) \, dx,$$

the mean $\bar{x}$ by

$$\bar{x} = \frac{1}{n} \int x . f(x) \, dx,$$

and the variance s^2 by

$$s^2 = \frac{1}{n}\int (x - \bar{x})^2 . f(x)\, dx,$$

$$= \frac{1}{n}\int x^2 . f(x)\, dx - \frac{1}{n}.2\bar{x}\int x . f(x)\, dx + \frac{1}{n}.\bar{x}^2 \int f(x)\, dx,$$

$$= \frac{1}{n}\int x^2 . f(x)\, dx - 2\bar{x}^2 + \bar{x}^2,$$

$$= \frac{1}{n}\int x^2 . f(x)\, dx - \bar{x}^2,$$

the integrals being taken over the whole range of possible values of the variable.

These formulae are the limiting forms of those given on pp. 14–17, namely

$$n = \Sigma f, \quad \bar{X} = \frac{\Sigma f X}{n} \quad \text{and} \quad s^2 = \frac{\Sigma f X^2}{n} - \bar{X}^2.$$

Example 5.4

A continuous variable is distributed from 0 to 4 with frequency-density given by $y = x^2(4 - x)$. Find the mean and the standard deviation.

$$\text{Total frequency} = \int_0^4 (4x^2 - x^3)\, dx, \quad = \tfrac{64}{3}.$$

$$\bar{x} = \tfrac{3}{64}\int_0^4 (4x^3 - x^4)\, dx, \quad = 2 \cdot 4.$$

$$s^2 = \tfrac{3}{64}\int_0^4 (4x^4 - x^5)\, dx - 2 \cdot 4^2, \quad = 0 \cdot 64.$$

$$\therefore \text{S.D.} = 0 \cdot 8.$$

The uniform distribution

This, the simplest of all cases, is of some theoretical importance. Suppose that a variable x is distributed uniformly from $-a$ to a with uniform density k. The mean is then zero, and the variance is

$$\frac{1}{2ak}\int_{-a}^{a} kx^2\, dx, \quad \text{i.e. } a^2/3.$$

The standard deviation is therefore $a/\sqrt{3}$. (This is analogous to the radius of gyration of a uniform rod of length 2a about an axis through its mid-point at right angles to its length.)

Probability density functions

If a member of the population is chosen at random, the probability for any interval δx of the variable is proportional to the frequency. The frequency curve thus becomes a probability curve, the scale being altered so that the total area under the curve is 1.

Let $p(x)$ be the probability-density at x. (It is assumed that $p(x)$ is finite.) Since the total probability is 1,

$$\int p(x)\,dx = 1,$$

the integral being taken over the whole range of possible values. If we assign zero probability to any intervals outside the range, we may write

$$\int_{-\infty}^{\infty} p(x)\,dx = 1.$$

The expected value of x is given by

$$E(x) = \mu = \int_{-\infty}^{\infty} x.p(x)\,dx;$$

and the variance by

$$\sigma^2 = \int_{-\infty}^{\infty} (x - \mu)^2.p(x)\,dx,$$

$$= \int_{-\infty}^{\infty} x^2.p(x)\,dx - \int_{-\infty}^{\infty} 2\mu x.p(x)\,dx + \int_{-\infty}^{\infty} \mu^2.p(x)\,dx,$$

$$= \int_{-\infty}^{\infty} x^2.p(x)\,dx - 2\mu^2 + \mu^2,$$

$$= \int_{-\infty}^{\infty} x^2.p(x)\,dx - \mu^2.$$

These results are analogous to those on pp. 56, 57, and might in fact have been deduced from them.

Independent continuous variates

It may well be inferred that the results of pp. 57–58 apply equally when the distributions are continuous.

*For a direct proof, suppose that x and y are continuous variates having distributions represented by the probability-density functions $p(x)$ and $q(y)$ respectively. Then, if x and y are independent, the probability that x takes a value between X and $X + \delta X$, and y takes a value between Y and $Y + \delta Y$ is $p(X)q(Y)\,\delta X\,\delta Y$.
Then

$$E(x+y) = \iint (x+y)p(x)q(y)\,dx\,dy$$
$$= \iint x.p(x)q(y)\,dx\,dy + \iint y.p(x)q(y)\,dx\,dy$$
$$= \int E(x).q(y)\,dy + \int E(y).p(x)\,dx$$
$$= E(x) + E(y).$$

and
$$E(xy) = \iint xy.p(x)q(y)\,dx\,dy$$
$$= \int x.p(x)\,dx . \int y.q(y)\,dy$$
$$= E(x).E(y).$$

For the variance of $(x+y)$, suppose that x and y are measured from their respective means. Then

$$E(x+y) = 0$$

and
$$\text{variance} = E(x+y)^2$$
$$= E(x^2) + E(2xy) + E(y^2)$$
$$= \sigma_x^2 + \sigma_y^2,$$

since $E(xy) = E(x).E(y) = 0$.
For the distribution of the mean of a sample the proof on p. 59 still applies.

Application

The algebraic types of distribution are chiefly of importance when considered as probability distributions. For example, when a number is given to two places of decimals there is an error that lies between ± 0.005. It is usually reasonable to suppose that all values within that range are equally likely. Hence the mean error is zero, and the variance is $(0.005)^2/3$, and the standard deviation is $0.005/\sqrt{3}$.
If n such numbers are added together, the variance of the total will

be n times as much, and the standard deviation $\sqrt{n}$ times as much. If the average of n such numbers is taken, the variance of the result will be

$$n \times \frac{1}{3}\left(\frac{0\cdot005}{n}\right)^2, \quad \text{or} \quad \frac{(0\cdot005)^2}{3n},$$

and the standard deviation $0\cdot005/\sqrt{(3n)}$. The error in the result is thus unlikely to be more than twice this amount, and very unlikely to be more than three times as much.

*Shepherd's adjustment

A closely related question is that of the error in a variance due to the grouping of the observations. If the group interval is c, the grouped readings may be regarded as the original readings to which have been added quantities varying between $-c/2$ and $+c/2$ and equally likely to be in any part of that range. The variance has thus been increased (on the average) by $c^2/12$. The correction for grouping is thus always negative and equal to $-c^2/12$. In Example 2.2, p. 18, this is $-1/12$, on the x-scale, and the corrected variance is therefore 4·263 approx. The corrected standard deviation, on the original scale, is thus $5 \times 2\cdot065$, or 10·3 years.

Attention should be given to the supposition made above that the effect of grouping is to add to each individual reading a quantity equally likely to have any value between $-c/2$ and $+c/2$. It might be thought that in a typical unimodal distribution this would not be so; but the point is that the boundaries between the groups are as likely to fall in one place as another. Shepherd's adjustment is an average adjustment for the different possible positions of the group boundaries. As will be shown in Exercise 6, No. 11, it does not necessarily improve the accuracy in each individual case. The conclusion from the study of such an example is that it is not worth while to apply Shepherd's adjustment unless the total frequency is very large and the group interval is a fairly considerable fraction of the standard deviation.

EXERCISE 6

1. A continuous variable is distributed between the values -1 and $+1$ with frequency-density given by $y = 60(1 - x^2)$. Find the total frequency and the standard deviation.
2. A continuous variable is distributed from 0 to 10 with frequency-density given by $y = 10x^2 - x^3$. Sketch the frequency curve and calculate the mean and standard deviation.

70   Statistics: the how and the why

3. A continuous variable is distributed with frequency-density y, where
$$y = 24x \quad \text{when} \quad 0 \leqslant x \leqslant 2,$$
and $\quad y = 60 - 6x \quad \text{when} \quad 2 \leqslant x \leqslant 10.$
Find the mean value of x and the standard deviation.

4. A variable x can take any positive value, the probability-density being proportional to e^{-kx}. Find in terms of k the expected value and the standard deviation. Find, also, the probability of a value less than the expected value.

5. Each of a series of observations is subject to an error x, varying from -2 to $+2$ units, the probability-density diminishing uniformly from $\frac{1}{2}$ at $x = 0$ to zero at $x = \pm 2$. Find the variance of the error and show that, if an average of ten such observations is taken, the error in the result is most unlikely to exceed 0·775.

6. An average of 50 readings is taken, each reading being correct to 3 places of decimals. Would the average be reliable to 4 places?

7. Thirty sums of money are each taken to the nearest £. Say what you can about the accuracy of (i) the total, (ii) the average.

8. A large number of sticks, whose lengths are uniformly distributed between 2·5 m and 3 m, are cut down to 2·5 m each. Show that, for a batch of 100 sticks, the total amount cut off will almost certainly lie between 20·7 m and 29·3 m.

9. A commuter has to change trains at three stations in the course of his daily journey. At the first of these he may have to wait for anything up to 8 minutes, at the second for anything up to 5 minutes, and at the third for anything up to 6 minutes. Find the mean total time spent waiting in the course of a journey, and the standard deviation. Assuming that the usual rough rule applies, determine 95 per cent confidence limits for the total time.

10. Two houses on a certain road are known to be situated one between the 9th and 10th milestones, the other between the 12th and 13th. Assuming they are equally likely to be at any point within these ranges, find the expected distance between them and the standard deviation.

11. The sales of footwear in a store in 52 successive weeks were as follows:
    37 40 46 47 47 49 51 53 54 54 56 56 57
    58 59 60 60 61 62 62 63 64 66 66 66 67
    67 68 68 68 69 69 70 70 70 71 73 73 74
    76 77 77 78 79 80 83 87 90 93 94 98 98
(The figures are arranged in order of magnitude for convenience.) Make frequency tables with class intervals 5, varying the position of the class boundaries. Find the variance in each case and apply Shepherd's adjustment.

(This is intended as an example for cooperative work. Individual students who do not wish to go through the full calculations should study the results as given in the Answers.)

6
The binomial distribution

Testing a hypothesis by repeated trials

The binomial distribution (p. 45) gives the probabilities of the different results in a series of repeated trials. For example, in tossing a coin, we might enquire whether a result of 4 heads in 4 throws is more remarkable than 8 in 10 throws. The probability of the first is $\frac{1}{16}$; and the probability of exactly 8 heads out of 10 is $_{10}C_2 \cdot (\frac{1}{2})^8 (\frac{1}{2})^2$, i.e. $\frac{45}{1024}$. But in judging which event is the more noteworthy it is of more interest to compare 4 out of 4 with 8 *or more* out of 10, for which the probability is $(45 + 10 + 1)/1024$, or $\frac{7}{128}$. This is just less than $\frac{1}{16}$. So 4 out of 4 is slightly more probable than 8 or more out of 10; and we can say that 8 out of 10 is the more remarkable of the two results.

If a man dies three days after his 100th birthday, there is not much point in enquiring what proportion of the population die at precisely that age (100 years and 3 days). It is of more interest to know how many people live beyond that age.

The following example may also help to make the point clear:

Example 6.1
Two helmsmen A and B race against each other on five successive days, drawing lots for boats on each occasion, and A wins 3 races out of 5 (dead heats being out of the question). Does this show that A is the more skilful?

We begin by supposing, as a null hypothesis, that they are equally skilful and that elements of luck are as likely to favour one as the other. Common sense tells us that one of them is bound, even so, to win at least three races; so the result provides no evidence at all of A's superiority. The chance that A should win exactly three times is $10/2^5$, or $\frac{5}{16}$; but the chance that he should win *at least* three times is $(10 + 5 + 1)/2^5$, or $\frac{1}{2}$ (in agreement with the common sense answer).

If A wins 4 times out of 5, the probability that this should happen by luck is $(5 + 1)/2^5$, or $\frac{3}{16}$, which is still by no means conclusive. But if he wins 8 times out of 10, the chance of his getting such a good result by luck is only $(45 + 10 + 1)/2^{10}$, or $\frac{7}{128}$. This is much stronger evidence of his superior skill.

Probability diagrams

The distribution of probabilities, in such examples as the above, can be represented diagrammatically. It is convenient to use the histogram form with an area scale for probability. Figure 6.1a shows such

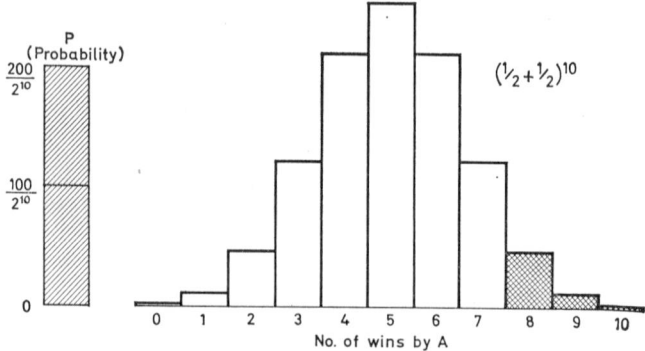

Fig. 6.1a. Sailing races

a diagram for the helmsmen when 10 races are sailed. The areas of the rectangles represent the terms in the expansion of $(\frac{1}{2} + \frac{1}{2})^{10}$ and the shaded part represents the probability of A winning at least 8 times. The total area is, of course, 1. The chance that they should each win 5 times is represented by the area of the tallest column, and it will be noted that, although 5 wins for each is more likely than any other result, it is not in itself very probable. (In fact, the probability is $\frac{252}{1024}$, or about 1 in 4.)

As the number of trials is increased, the chance of an equal division becomes less: for 20 trials it is about 1 in 5·7, for 40 about 1 in 8·2. This is shown in Fig. 6.2a, p. 74.

If the chances on a single trial are not equal, the binomial distribution is unsymmetrical. If an ordinary die is thrown 10 times, and the throwing of a 'six' is counted a 'success', the probabilities of 0, 1, 2, 3, ..., 10 successes are given by the terms in the expansion of $(\frac{5}{6} + \frac{1}{6})^{10}$, as shown in Fig. 6.1b. In this distribution the highest

The binomial distribution

column indicates the most probable number of successes, namely 1. This corresponds to the mode of a frequency distribution. Corresponding to the mean we have the 'expected number of successes', $1\frac{2}{3}$ (as will be proved later).

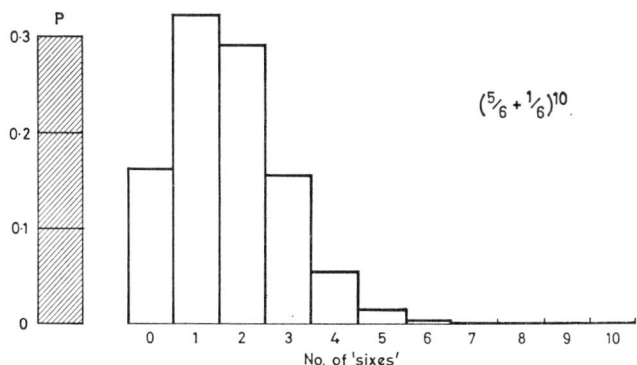

Fig. 6.1b. Throwing a die

If the number of trials is at all large, it is not practicable to calculate the probability of any particular number of successes being exceeded. In such a case it is useful to know the standard deviation of the distribution, so that we may use the rough rule that about 5 per cent of the area lies outside the range $m \pm 2s$.† We therefore proceed to find the mean and the standard deviation of the probability distribution represented by the terms of the expansion of $(q + p)^n$. The calculation follows the same method as for frequency distributions, but it is convenient to arrange the terms in rows rather than in columns.

x (no. of successes)	0	1	2	3	...	n
P (probability)	q^n	$nq^{n-1}p$	$\frac{n(n-1)}{1.2}q^{n-2}p^2$	$\frac{n(n-1)(n-2)}{1.2.3}q^{n-3}p^3$	...	p^n
Px	0	$nq^{n-1}p$	$\frac{n(n-1)}{1}q^{n-2}p^2$	$\frac{n(n-1)(n-2)}{1.2}q^{n-3}p^3$	...	np^n
Px^2	0	$nq^{n-1}p$	$\frac{2n(n-1)}{1}q^{n-2}p^2$	$\frac{3n(n-1)(n-2)}{1.2}q^{n-3}p^3$	...	n^2p^n

† This rule is not applicable unless the distribution is reasonably symmetrical. It should not therefore be used if p is near 0 or 1.

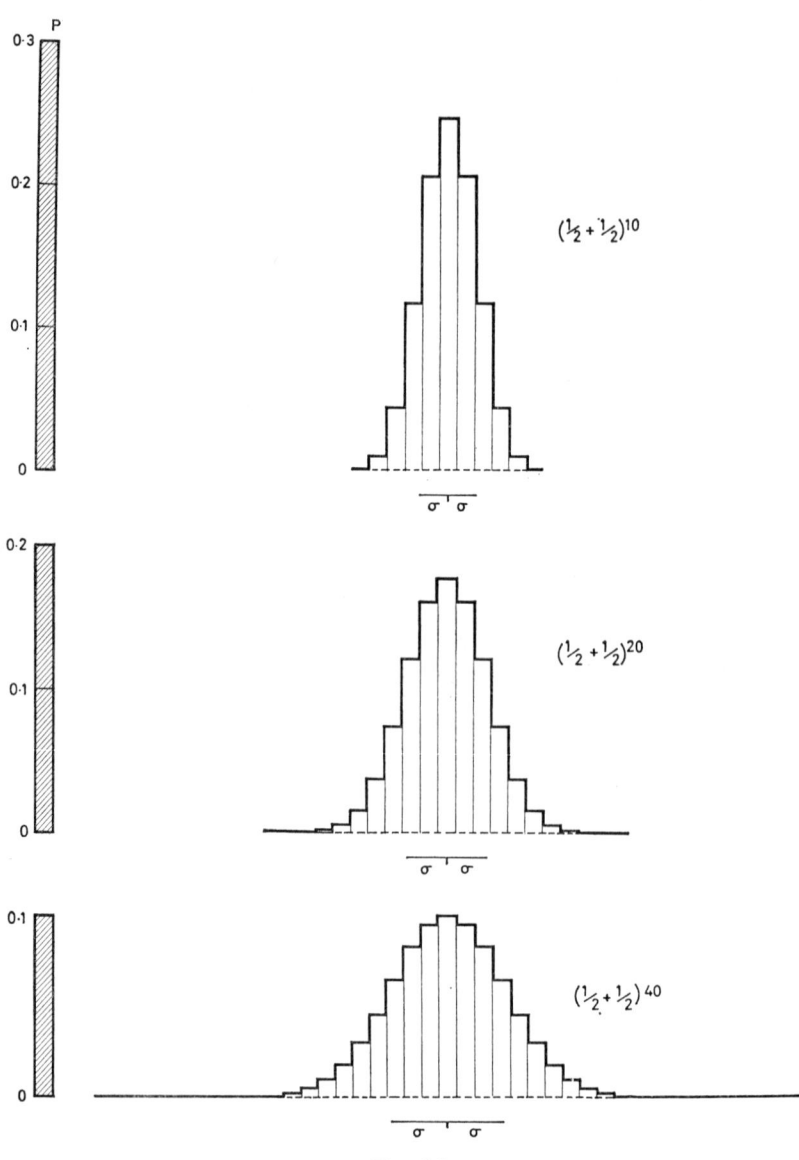

Fig. 6.2a

$$\sum Px = np\left\{q^{n-1}+(n-1)q^{n-2}p+\frac{(n-1)(n-2)}{1.2}q^{n-3}p^2+\cdots+p^{n-1}\right\},$$
$$= np(q+p)^{n-1},$$
$$= np.$$

Since the total probability is 1, the mean, or 'expected value' is np.

$$\sum Px^2 = \sum Px$$
$$+\left[\frac{n(n-1)}{1}q^{n-2}p^2+\frac{2n(n-1)(n-2)}{1.2}q^{n-3}p^3+\cdots+(n^2-n)p^n\right],$$
$$= np + n(n-1)p^2(q+p)^{n-2},$$
$$= np(1 + np - p).$$

For the variance we divide by the total, 1, and subtract the square of the mean, n^2p^2. Thus:

$$\text{variance} = np(1-p) = npq,$$
and $$\text{standard deviation} = \sqrt{(npq)}.$$

If a coin is tossed 100 times, the expected number of heads is 50 and the standard deviation is $\sqrt{(100 \times \frac{1}{2} \times \frac{1}{2})}$, i.e. 5. We should, therefore, infer a 95 per cent probability that the number of heads would lie between 40 and 60, and a very small probability indeed that it would be outside the range 35 to 65.

It is apparent from the formula for the standard deviation that, as n increases, the probable spread away from the expected value also increases, but not as fast as n (Fig. 6.2a). This suggests consideration of the proportion (or percentage) of successes in n trials. Dividing the previous results by n, we see that the expected proportion of successes is p, with a standard deviation of $\sqrt{(pq/n)}$. For the coin tossed 100 times this gives an expected proportion of 0·5 (50 per cent) with standard deviation 0·05 (5 per cent), in agreement with the previous results. The formula shows that, as n increases, the percentage spread away from the expected value diminishes (Fig. 6.2b). This illustrates *Bernoulli's Theorem*, that, as n increases, the probability of the proportion of successes being within a fixed range $p \pm \epsilon$ also increases.

In Fig. 6.2a, the length of the base increases with n, the height of the centre decreasing, and the standard deviation increasing as $\sqrt{n}$. In Fig. 6.2b, the 'horizontal' scale is reduced as n increases, in the ratio $1/n$, so that the length of the base is constant. To keep the

76 Statistics: the how and the why

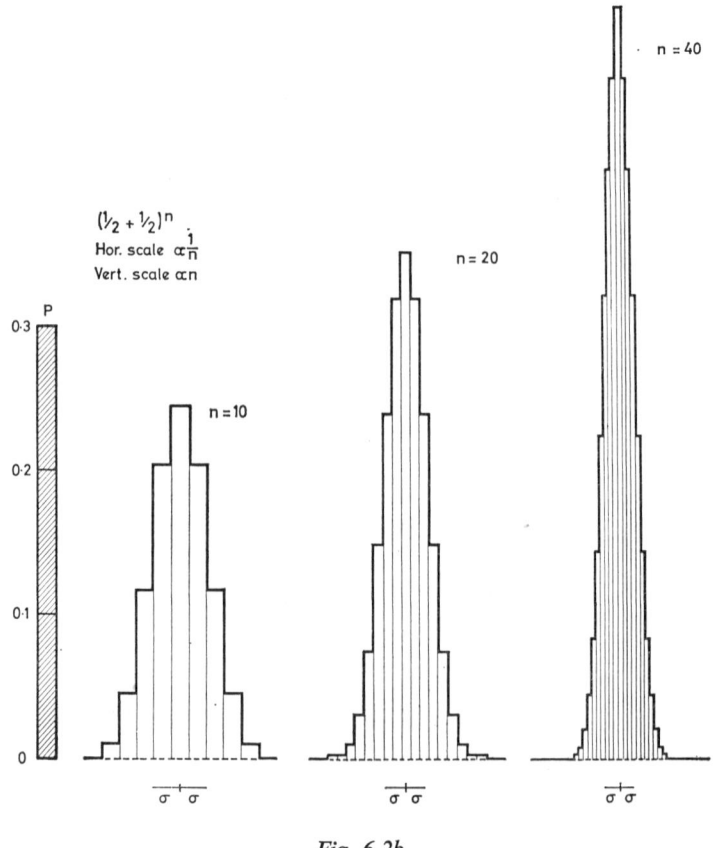

Fig. 6.2b

area 1 unit the 'vertical' scale is increased in the ratio $n/1$, and the standard deviation now diminishes as n increases, according to the formula

$$\sqrt{\left(\frac{1}{2} \times \frac{1}{2} \times \frac{1}{n}\right)}.$$

* It may occur to some readers that there should be a connection between the above results and the formula $\sigma/\sqrt{n}$ for the standard deviation of the mean of a sample of n members. It is true that a series of n trials can be regarded as a sample of n observations; but it must be remembered that σ (in the formula $\sigma/\sqrt{n}$) is the standard deviation for the probability distribution of a single observation. The probabilities for a single trial are as follows:

No. of successes	0	1
Probabilities	q	p

Applying the usual method, the mean, or expected value is p, and the variance is $p - p^2$, i.e. pq. Hence, $\sigma = \sqrt{(pq)}$, and $\sigma/\sqrt{n} = \sqrt{(pq/n)}$.

* For a proof of Bernoulli's Theorem it is necessary first to prove Tchebycheff's Theorem, that the chance of a deviation from the expected value exceeding $\lambda\sigma$ is less than $1/\lambda^2$.

Since σ^2 is the sum of the squares of the deviations each multiplied by its probability, it follows that, if P is the probability of a deviation greater than $\lambda\sigma$, $\sigma^2 > P\lambda^2\sigma^2$ (Fig. 6.3). Therefore, $P < 1/\lambda^2$.

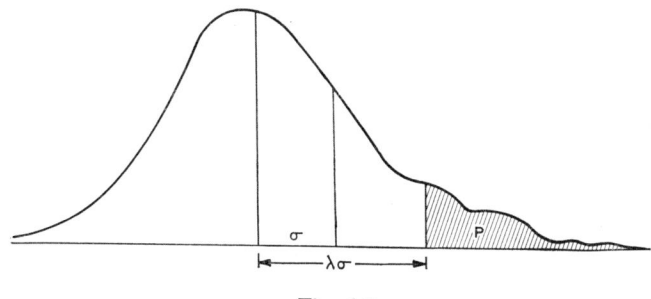

Fig. 6.3

For a given deviation ϵ, equal to $\lambda\sigma$, $P < \sigma^2/\epsilon^2$. In the case of the binomial distribution, with ϵ measured as a percentage deviation, $\sigma^2 = pq/n$, and, therefore, $P < pq/n\epsilon^2$, which diminishes as n increases, approaching zero as $n \to \infty$.

Applications

The binomial distribution is applicable only when the probability of success is the same in each trial. This implies that the trials must be independent of one another.

If, for example, cards are drawn, one from each of two packs, the chance of a 'snap' is $\frac{1}{52}$. But if a second pair of cards is then drawn, without the first pair being replaced, the chance is not the same. If, however, n pairs are drawn in succession, each pair being replaced and the packs shuffled before the next pair is drawn, the probabilities of there being 0, 1, 2, ..., n 'snaps' are given by the terms in the expansion of

$$\left(\frac{51}{52} + \frac{1}{52}\right)^n.$$

EXERCISE 7

1. Six cards are drawn in succession from an ordinary pack, each being replaced and the pack shuffled before the next is drawn. Find the probabilities:
   (i) that three should be red and three black,
   (ii) that at least four should be spades,
   (iii) that not more than four should be spades.
2. If 5 per cent of a consignment of articles are defective, what is the chance that a sample of 10 would not contain any defectives? What is the chance that it would not contain more than one?
3. If, in the works of a certain author, one letter in 9 is an E, show that the probability that 50 letters picked at random should not include an E is about 1 in 360. Express, similarly, the probability that they should not include more than one E.
4. If one letter in 9 is an E, what is the expected number of Es in a passage containing 200 letters? A count of six such passages by different authors gave the following results: 29, 23, 20, 17, 24, 20. Would you regard any of these as exceptional on the 1 in 9 hypothesis?
5. The French naturalist Buffon tossed a coin 4040 times, obtaining 1992 tails to 2048 heads. Was this result at all surprising?
6. A die was thrown 480 times and the number 3 turned up 108 times. Would you suspect that the die was biased?
   In a second experiment with the same die, the number 3 turned up 103 times out of 420 throws. Does this strengthen the impression?
7. The Swiss astronomer Wolf threw a pair of dice 100,000 times and obtained a proportion of 0·83533 unlike pairs. Find the excess above the expected result and find whether it was unusually large.
8. In an experiment on telepathy, a pack of 25 cards of five kinds was used. A card was drawn from the pack and a person in another room had to say which of the five kinds it was. In 800 trials he was right 207 times. Would you regard this as a significant result?
9. A large batch of manufactured articles is accepted if
   either (*a*) a random sample of 6 articles contains not more than one defective article,
   or (*b*) a random sample of 6 contains two defectives and a second random sample of 6 contains none.

If, in fact, 20 per cent of the articles in the batch are defective, what is the chance of the batch being accepted?

10. A correspondent wrote to *The Times* that in 14 years he had recorded rain on 2248 days, of which 350 were Thursdays. Is this evidence that Thursday tends to be a rainy day?

11. Just before the 1964 election the weekly National Opinion Polls were based on samples of 2000 voters who were usually almost equally divided between the two main parties. What would be the least percentage change in the results that could reasonably be taken as indicating a swing of voting opinion?

12. In a study on bronchitis among forestry workers† it was found that in the UK as a whole the prevalence, in one year, was 4·91 per 100 workers. When the workers were divided according to whether their homes were in rural or non-rural areas, the results were as follows:

	No. of workers	Spells of bronchitis	Prevalence per 100
Non-rural	1476	83	5·62
Rural	8382	400	4·77
Total	10832	532	4·91

(The total included some districts in which the workers were not classified.)

On the assumption that place of residence does not affect the risk of bronchitis, determine the 'expected' number of spells for the non-rural group, and find whether the observed number is significantly different. Show that the same result may be obtained by using the formula $\sqrt{(pq/n)}$ on the figures in the last column.

13. In the 1961 census in England and Wales, supplementary questions were asked of one household in 10; but it was afterwards discovered that this sample contained a 10 per cent deficiency of one-person households, only 3600 such households being considered in a district containing 40,000. Was the sampling biased?

14. An advertiser sends out circulars in batches and finds that, on the average, one circular in five produces a reply. If he sends out a batch of 100 and gets 25 replies, would you consider this result to be significantly above average? If he then sends out a further 500 of the same circular and gets 120 replies, find whether the combined result for the 600 is significantly above average.

15. In a certain plant 75 per cent of specimens show a particular feature. State, with reason, how large a sample should be taken so that there will be only 1 chance in 20 of the percentage observed being outside the limits 70 per cent to 80 per cent.

16. Samples, each of 8 articles, are taken at random from a large consignment in which 20 per cent of the articles are defective. Find the

† Data from Dr D. McGregor.

most likely number of defective articles in a single sample, and the chance of obtaining precisely this number.

If 100 samples of 8 are to be examined, calculate the number of samples in which you would expect to find three or more defective articles.

17. A large batch of manufactured articles is accepted if either of the following conditions is satisfied:
    (a) A random sample of 10 articles contains no defectives.
    (b) A random sample of 10 contains one defective, and a second random sample of 10 contains none.

    If, in fact, 5 per cent of the articles in a batch to be examined are defective, find the chance of the batch being accepted.
    $$[(0 \cdot 95)^{10} = 0 \cdot 5987.]$$

18. When a stick of length l is thrown at random on to a plane covered by parallel lines spaced at intervals l, the chance that it will fall across one of the lines is $2/\pi$, or $0 \cdot 637$. (At an angle θ to the lines, the chance is $\sin \theta$, and the mean value of this, for $0 < \theta < \pi/2$, is $(2/\pi) \int_0^{\pi/2} \sin \theta \, d\theta$.)

    Five boys carried out an experiment, throwing the stick 405 times in all and obtaining a result of $0 \cdot 656$. Is this reasonably near the expected value? One of them, who threw it 60 times, obtained $0 \cdot 73$. Is there any reason to suppose that he failed to throw it in a random manner?

19. How many times would the stick in Q. 18 have to be thrown to give a 95 per cent chance of obtaining a value of $2/\pi$ correct to three figures?

7
The Poisson distribution

It often happens that the probability of an event is small, but the number of 'trials' is large. The chance that any particular person should contract appendicitis in a specified week is a very small one, but owing to the fact that there are a very large number of people at risk, the hospitals do in fact receive a fairly steady stream of cases. Again, the chance that a particular telephone subscriber in one town should want to ring a number in a neighbouring town during a five-minute period is a small one, but because there are thousands of subscribers, the lines between the two towns may have to carry several calls at once. If the average number of simultaneous calls is 5, the telephone engineers will want to know the probability of 6, 7, 8 or higher numbers of calls occurring at the same time.

It is, therefore, of interest to know what happens to the binomial distribution when p is small, n large, and the mean value np has a finite value m. As with the binomial distribution, it is necessary to suppose that the happenings occur at random and independently of each other. Thus it is supposed that appendicitis is neither a seasonal disease (like hayfever) nor infectious (like measles). In the case of the telephone calls, our conclusions would be affected if a considerable number of people were in the habit of putting through calls at regular intervals, or if some special event, such as a test match or an earthquake, caused a large number of anxious enquiries to be made at the same time.

In the binomial distribution the first term, q^n, represents the probability of 0 successes. We require an approximate value for it when p is small, n is large, and $np = m$.

$$q^n = (1-p)^n = \left(1 - \frac{m}{n}\right)^n,$$

and it is proved in books on analysis that, as n increases, m being constant, this expression approaches the value e^{-m}. The following table shows that e^{-m} is quite a good approximation when $p \ (=m/n)$ is as small as 0·02, provided m is not too large:

Values of $(1 - m/n)^n$

$p \backslash m$	2	5	10	20
0·1	0·1215	0·0051	0·000026	$10^{-9} \times 0·705$
0·05	0·1285	0·0059	0·000035	$10^{-9} \times 1·228$
0·02	0·1326	0·0064	0·000041	$10^{-9} \times 1·683$
0·01	0·1340	0·0066	0·000043	$10^{-9} \times 1·864$
e^{-m}	0·1353	0·0067	0·000045	$10^{-9} \times 2·064$

The second term of the binomial distribution is $nq^{n-1}p$, i.e. $np/q \times$ the first term. Its approximate value is therefore me^{-m}. The third term is obtained from the second by multiplying by

$$\frac{n-1}{2} \cdot \frac{p}{q}, \quad \text{or} \quad \frac{m}{2}\left(1 - \frac{1}{n}\right).$$

Its approximate value, when n is large, is therefore $(m^2/2)e^{-m}$; and so on. So the first few terms are approximately the terms of the series

$$e^{-m}\left\{1 + m + \frac{m^2}{2!} + \frac{m^3}{3!} + \cdots\right\}.$$

These terms represent approximately the probabilities of $0, 1, 2, 3, \ldots$, successes when n (the number of 'trials') is large and p (the probability of success in any one trial) is small, and $np = m$.† The distribution is named after S. D. Poisson (1781–1840), who wrote on many branches of applied mathematics, including probability.

Generally speaking, it is only the first few terms that are needed, but it will be noted that the sum of the whole series is 1. To find the expected number of successes we multiply the terms by $0, 1, 2, 3, \ldots$, respectively, obtaining

$$\mu = e^{-m}\left\{0 + m + m^2 + \frac{m^3}{2!} + \cdots\right\},$$

and this is equal to $e^{-m} \cdot me^m$, i.e. m.

For the variance, we multiply again by $0, 1, 2, 3, \ldots$, and, after summing, subtract m^2. This gives

$$\sigma^2 = e^{-m}\left(0 + m + 2m^2 + \frac{3m^3}{2!} + \cdots\right) - m^2.$$

† It can be proved that these terms are the limiting values of the binomial terms as $n \to \infty$, with $np \ (=m)$ constant. But in applications n is large, not infinite.

This exceeds the mean by

$$e^{-m}\left(m^2 + \frac{2m^3}{2!} + \frac{3m^4}{3!} + \cdots\right) - m^2,$$

or $\quad e^{-m}.m^2\left(1 + m + \frac{m^2}{2!} + \cdots\right) - m^2.$

It is therefore equal to $m + m^2 - m^2$, i.e. m.

Thus the variance of this distribution is equal to the mean.

Example 7.1

In an examination room there were 88 *boys and the number of coughs each minute during a half-hour period were recorded as follows:*

No. of coughs	0	1	2	3	4	5	6	7
Frequency	5	10	5	7	2	1	0	0

Total 30.

The mean number of coughs per minute is found to be 1·8 and the variance 1·76. (This close agreement does not, of course, prove that the distribution is Poissonian, but at least it does not disprove it.) Supposing, as a null hypothesis, that the coughs were independent and that 88 was a sufficiently large value of n, the probabilities of the different numbers of coughs in any minute would be given by the terms of

$$e^{-1\cdot 8}\left(1 + 1\cdot 8 + \frac{1\cdot 8^2}{2!} + \frac{1\cdot 8^3}{3!} + \cdots\right),$$

and the expected frequencies of 0, 1, 2, 3, ..., coughs, during the half-hour, would be given by

$$30e^{-1\cdot 8}\left(1 + 1\cdot 8 + \frac{1\cdot 8^2}{2!} + \frac{1\cdot 8^3}{3!} + \cdots\right).$$

The values so obtained are

    4·96, 8·93, 8·04, 4·82, 2·17, 0·78, 0·23, 0·06, ...

Note. It may be objected that in this and similar examples the probability p is not constant, but varies from one individual to another. This, however, does not matter because, as will be proved later (p. 86), the sum of any number of independent Poisson variates with different values of m is also a Poisson variate. If, in the above example, the coughs of each individual boy are assumed to be at random and, hence, that their probability distribution is Poissonian, it follows, even though some may cough more than others, that the total numbers of coughs will have a probability distribution of the same kind.

Example 7.2 (Quality control)
In a mass-production process, 5 per cent of the products are, on the average, defective (in the sense of failing to pass a certain test). If samples of 10 are tested the 'expected number' of defectives in a sample is 0·5, and the probabilities of 0, 1, 2, 3, ..., defectives are, therefore, given by the terms of

$$e^{-0.5}\left(1 + 0.5 + \frac{0.5^2}{2!} + \frac{0.5^3}{3!} + \cdots\right),$$

i.e. $\quad 0·607 + 0·303 + 0·076 + 0·013 + 0·002 + \cdots$

(It might be thought that $p = 0·05$ is not a sufficiently small value, nor $n = 10$ a sufficiently large one, for the distribution to be counted as Poissonian. The corresponding binomial values are

$$0·599 + 0·315 + 0·075 + 0·010 + 0·001 + \cdots$$

and it will be seen that the agreement is sufficiently close.)

Using the Poisson values, the chance of there being two or more defectives is $1 - 0·607 - 0·303$, or 0·09; and the chance of 3 or more is similarly 0·014. If the sample contained 3 defectives, suspicion would be aroused that the 5 per cent average was being exceeded; and 4 defectives would make this almost a certainty.

The record of this sampling might be kept in the form of a control chart, as shown in Fig. 7.1.

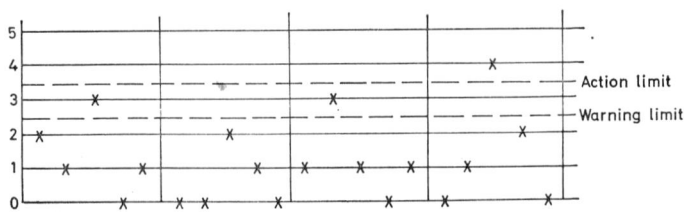

Fig. 7.1. Control chart

Note. In the above example, the values of n and p were known and, hence, m ($=np$) was found. In Example 7.1, a value for m was found by observation. (n was in fact known to be 88, but p was unknown.) There are many other cases (e.g. Exercise 8, No. 1) in which a value can be found for m although n and p are quite unknown, except in so far as n is known to be large and p small.

EXERCISE 8

1. The number of road accidents notified to a certain police station per day is shown in the following frequency table relating to a period of 300 successive days:

Accidents per day	0	1	2	3	4	5	6	7
Frequency	90	113	64	21	7	3	1	1

    Calculate the mean number of accidents a day. Use the Poisson distribution, with this mean, to calculate the 'expected' frequencies. Assuming that this distribution continues to apply, find the chance of four or more accidents being notified on any one day.

2. In 90 consecutive issues of *The Times* a count was made of the numbers of births with initial letters A and B. The frequencies with which the different numbers of such births occurred were as follows:

No. of births recorded	Frequencies A	Frequencies B
0	37	11
1	35	24
2	14	22
3	3	21
4	1	7
5	—	4
6	—	1
	90	90

    Find the mean number per issue for each of these letters, and the expected frequencies, assuming a Poisson distribution.

3. At a telephone exchange, during a certain part of the day there are, on the average, 4 calls a minute. How often, during a period of 60 consecutive minutes, would you expect there to be **(i)** no calls during the minute, **(ii)** more than 5 calls?

4. In 300 shifts in a factory 60 accidents occurred, as shown in the following table:

Accidents per shift	0	1	2	3	4
Frequency	248	45	6	1	0

    Assuming a Poisson probability distribution, find the probability of more than one accident occurring in a shift.

5. In a manufacturing process, 3 per cent of the products are defective (i.e. fail to reach a certain standard). Random samples of 10 are tested at intervals. Evaluate the probability that a sample should contain no defectives, and show that $e^{-0.3}$ gives a good approximation. Using the Poisson distribution as an approximation, find the probability that a sample should contain (a) more than one defective, (b) more than two. What action would you recommend if a sample were found to contain (i) two defectives, (ii) three defectives?

6. A process that normally produces 5 per cent of defective products is checked by periodical tests on samples of 10. What conclusions, if any, would you draw (i) if 5 consecutive samples were said to contain no defectives, (ii) if 10 consecutive samples were said to contain no defectives?

7. If the probability that a man aged 35 should die within a year is 0·0063, find the probabilities that, of a group of 100 men of that age, (i) all should survive the year, (ii) more than one should die.

8. A life assurance company found that, over a period of 4 years, 14·1 per cent of deaths among its policy holders were due to respiratory diseases, and 0·55 per cent to tuberculosis other than respiratory. In the following year, out of 1308 deaths, 125 were in the first category and only 1 in the second. Find whether these figures represent significant reductions.

9. During the years 1950–57 inclusive the mileage of live-rail electrified track on British Railways was approximately 2700. During those years, 43 child trespassers were killed by contact with live rails. If, during one year, 3 such separate casualties were to occur on separate occasions on a particular stretch of 100 miles, would you say that special protection was needed on that stretch of line?

10. A sample of 12 supposedly identical electronic components is tested, and none of them fails. Show that if, in fact, the reliability were 90 per cent (i.e. if there were a 10 per cent chance of any one of the 12 items failing) there would have been approximately a 72 per cent chance of a failure in the sample.

*The sum of two independent variates

A *variate* is a variable with which is associated a certain pattern of probabilities. A *Poisson variate with parameter m* is one for which the values 0, 1, 2, 3, ..., have probabilities

$$e^{-m}\left\{1, m, \frac{m^2}{2!}, \frac{m^3}{3!}, \ldots\right\}.$$

The sum of two independent Poisson variates with parameters m and n is a Poisson variate with parameter $(m + n)$.

Proof: If x and y are the variates, the probabilities of their taking the different values are as shown below:

x	0	1	2	3	...
x	$e^{-m}\{1$	m	$\dfrac{m^2}{2!}$	$\dfrac{m^3}{3!}$	$...\}$
y	$e^{-n}\{1$	n	$\dfrac{n^2}{2!}$	$\dfrac{n^3}{3!}$	$...\}$

The probability that $x + y = 0$ is
$$e^{-m}.e^{-n}, \quad \text{i.e. } e^{-(m+n)};$$
the probability that $x + y = 1$ is
$$e^{-(m+n)}(m + n);$$
the probability that $x + y = 2$ is
$$e^{-(m+n)}\left\{\frac{m^2}{2!} + mn + \frac{n^2}{2!}\right\}, \quad \text{i.e. } e^{-(m+n)}.\frac{(m+n)^2}{2!};$$
and, in general, the probability that $x + y = r$ is
$$e^{-(m+n)}\left\{\frac{m^r}{r!} + \frac{m^{r-1}}{(r-1)!}\frac{n}{1!} + \cdots + \frac{m^{r-k}}{(r-k)!}\frac{n^k}{k!} + \cdots + \frac{n^r}{r!}\right\},$$
i.e.
$$e^{-(m+n)}\frac{(m+n)^r}{r!}.$$

Therefore the probabilities for $(x + y)$ to take the values 0, 1, 2, 3,..., are given by
$$e^{-(m+n)}\left\{1, \ (m+n), \ \frac{(m+n)^2}{2!}, \ \frac{(m+n)^3}{3!}, \ \ldots\right\}.$$

The Poisson summation chart

In applications, we usually require the probability that the variate should be greater than or equal to a given value r. The chart† shown in Fig. 7.2 gives this for different values of m. To use it, find the value of m on the 'horizontal' axis; follow the ordinate at this point upwards till it meets the curve labelled with the given value of r; then the value on the 'vertical' scale gives the probability that the variate will have a value greater than or equal to r. For example, if the mean

† Copies (e.g. in the Chartwell series) may be obtained through any stationer.

88 Statistics: the how and the why

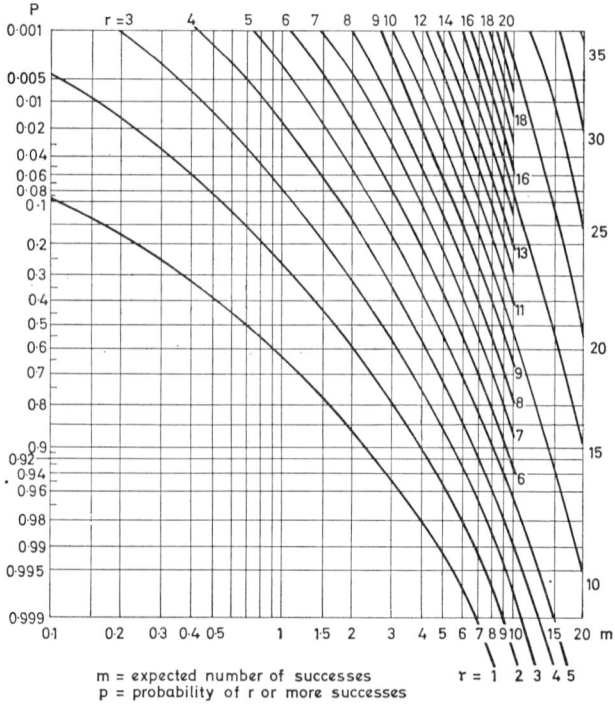

Fig. 7.2. The Poisson summation chart

value of a Poisson variate is 3, the probability that its value will be 2 or more is 0·8; and the probability that it will be 10 or more is 0·001.

Most of the questions in Exercise 8 can be done more quickly with the aid of this chart.

The chart may also be used to test whether a given set of readings follows approximately the Poisson pattern. It is necessary to tabulate readings for '1 or more', '2 or more', etc., and to express them as proportions of the total frequency. Points are then plotted on the curves $r = 1, r = 2$, etc., at the appropriate heights and, if the distribution is Poissonian, the points will lie on a 'vertical' line $x = m$.

EXERCISE 9

1. During a certain period of the day the cash desk of a store has to deal with an average of 5 transactions a minute. Find from the Poisson chart

(i) the probability that there should be 9 or more transactions in a particular minute,
(ii) the probability that there should be 3 or less.

2. At a telephone exchange there is, during a certain period, an average of 7 calls a minute. Find the probabilities that during a particular minute there should be
   (i) more than 10 calls,
   (ii) no calls at all,
   (iii) not more than one.

3. In a factory, electric lamps have to be replaced at an average rate of 15 a day. Find the chance that in one day there should be (i) not more than 9 replacements, (ii) 25 or more.

4. In England and Wales, 1 person in 73 is called Smith, and 1 in 174 is called Brown. In a school of 433 boys there were 3 Smiths and 4 Browns. Find the probabilities that there should be (i) 3 or less Smiths, (ii) 4 or more Browns.

   In the same school, 1 boy in 11 had a surname beginning with B, but a statistics class of 10 boys contained 5 with initial letter B. Estimate the probability of this happening by chance.

5. In 1956, the number of wage-earners employed in British coal mines was approximately 700,000. Accidents caused 318 deaths during the year, and 1731 reportable injuries. In the south-eastern division, in which the number of wage-earners was 7100, there were 5 deaths and 6 reportable injuries. Would you regard either of these figures as remarkably high or low?

6. A doctor finds that, among his patients, a certain condition occurs on the average in 10 cases a year. These are isolated cases, but the condition can also occur in epidemic form. Should he suspect an epidemic if he meets (a) 3 cases, (b) 4 cases, in a particular fortnight?

7. During 1945, an investigation was made by Mr R. D. Clarke to find whether the flying bombs, which were then falling on London, tended to fall in clusters. An area of South London was divided into 576 squares of $\frac{1}{4}$ sq. km each, and a count was made, with the following results:

No. of bombs per square	0	1	2	3	4	5
No. of squares	229	211	93	35	7	1

   Show, by plotting on a Poisson chart, that the distribution was approximately Poissonian, and estimate the mean number per square. Find the mean number also by calculation and use this value to calculate the expected frequencies.

8. In a manufacturing process the proportion of defective products is normally 32 per thousand. How large a sample should be taken so that the chance of 3 or more defectives in the sample would be about 1 per cent?

9. A tradesman sells washing machines, which are delivered by a wholesaler once a month. On the average, he sells 6 per month. How many should he have in stock just after a delivery for the chance of running short to be less than 1 in 10?

10. The full table for Exercise 8, No. 2, showing the distribution of initial letters A and B in births recorded in *The Times*, was as follows:

		B						
		0	1	2	3	4	5	6
A	0	4	9	9	9	4	2	–
	1	6	7	10	8	2	1	1
	2	1	6	3	4	–	–	–
	3	–	2	–	–	1	–	–
	4	–	–	–	–	–	1	–

Tabulate the frequencies for initials A and B together.

Make a table of cumulative frequencies showing values for '1 or more', '2 or more', etc.

Use a Poisson chart to show that the first six of these values indicate that the distribution is approximately Poissonian, and estimate the value of m. Compare this with the sum of the means obtained in Exercise 8, No. 2 for the separate letters. Calculate the expected frequencies for A and B together and plot a graph to compare them with the observed values.

8
The normal distribution

Limiting form of the binomial distribution

It is natural to enquire what happens to the binomial distribution when n is increased to a large number. One reason for this is that the applications are of interest. (What, for example, is the probable result of spinning a coin a very large number of times?)

The diagrams of Fig. 6.2a suggest that, as n is increased, the figure gradually flattens out, becoming more and more like a straight line with a slight hump in the middle. The length of the line represents n, and increases steadily. The width of the hump also increases, but not so rapidly. It is best measured by the standard deviation, which increases proportionally to $\sqrt{n}$.

If, to avoid losing the form of the curve altogether, we reduce the 'horizontal' scale in the ratio $1/n$, thus keeping the base-line of constant length, and at the same time increase the 'vertical' scale in the ratio $n/1$ (so as to keep the total area constant), we obtain the results shown in Fig. 6.2b (p. 76). Now, as n is increased, the peak rises and the standard deviation decreases, proportionally to $1/\sqrt{n}$. The diagram appears to be approaching the shape of an inverted T and its form is again lost.

An obvious remedy is to compromise by reducing the 'horizontal' scale in the ratio $1/\sqrt{n}$ instead of $1/n$. The standard deviation then remains constant, and this in itself is an advantage, since we may want to use the final result for comparison with observed frequency distributions whose standard deviation is known. The diagrams of Fig. 8.1 show that, under this system, the peak remains at a nearly constant height and that the form approaches that of a well-defined curve. This curve is known as *the normal curve of errors*, or more shortly as *the normal curve*. Its equation is

$$Y = \frac{1}{\sqrt{(2\pi\sigma^2)}} e^{-X^2/2\sigma^2},$$

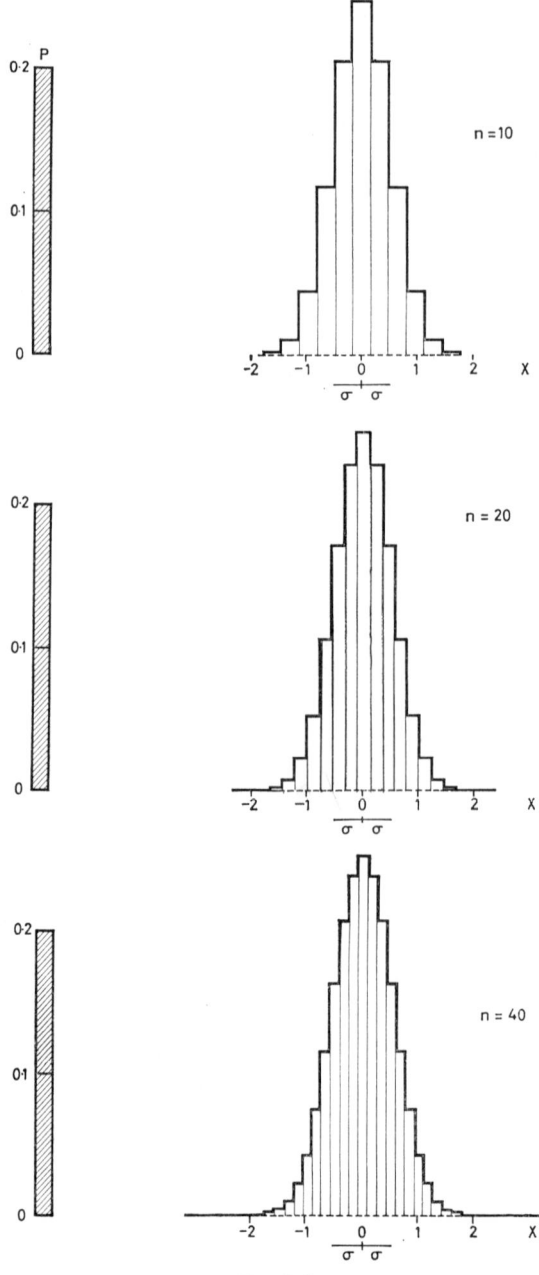

Fig. 8.1

$$p(x) = \frac{1}{\sqrt{(2\pi\sigma)}} e^{-\frac{x^2}{2\sigma^2}}$$

The normal distribution

where σ is the standard deviation and X is measured from the centre of the curve. (A proof of this will be found at the end of the chapter.)

A somewhat surprising fact is that the same form is arrived at even if the values of p and q are unequal, provided that neither of them is very small. The diagrams for $(\frac{4}{5} + \frac{1}{5})^n$ are shown in Fig. 8.2.

Interpretation of the normal curve

The diagrams of Fig. 6.2a had a definite meaning in terms of probability. The area of each rectangular space under one of the 'steps' represented the probability of obtaining a certain number of 'successes' in n trials. If P is this probability, $P = y\,\delta x$, where y is the height of the rectangle and δx the width, δx being equal to 1 unit. When we change the scale, $x = \sqrt{n}.X$, and $y = Y/\sqrt{n}$, so that P is represented by the same area as before. $\delta x = \sqrt{n}.\delta X$, so that, whereas δx was equal to 1, δX diminishes as n increases. The rectangles narrow and the diagram approaches the form of a continuous curve. The standard deviation is $\sqrt{(pq)}$ on the new scale, and remains constant.

The height of the curve at any point represents a 'probability-density'; that is to say that if δX is any small interval on the X-axis (not now limited to cover only one of the original rectangles) the probability of the number of 'successes' falling in the interval represented by δX is approximately $Y\,\delta X$.

It is also necessary to be clear about the meaning of the X-scale. The origin ($X = 0$) represents the 'expected' number of successes (as appears in the proof at the end of the chapter), and a point X represents the excess above that number, but on a reduced scale, the actual excess being $\sqrt{n}.X$. The quantity $X/\sqrt{(pq)}$ measures the excess in comparison with the standard deviation, being equal to $x/\sqrt{(npq)}$ on the old scale.

If an ordinate is drawn at any point X, the area under the curve to the right of that ordinate represents the probability of the value being exceeded. Tables of the area show values corresponding to various values of X/σ ($= X/\sqrt{(pq)}$). The table on p. 95 shows areas from 0 to X/σ, the whole area under the curve being, of course, 1 unit. Thus, to find the probability of the value being exceeded it is necessary to subtract the reading from 0·5 (or, if X is negative, to add 0·5).

It will be seen from the table that the probability of there being an excess above the expected value amounting to more than 2σ is 0·5 − 0·4772, i.e. 0·0228. Doubling this, the chance of a deviation either way amounting to more than 2σ is 0·0456, or about $4\frac{1}{2}$ per cent. This is in line with the rough rule given in Chapter 2 (p. 18) that in

94 Statistics: the how and the why

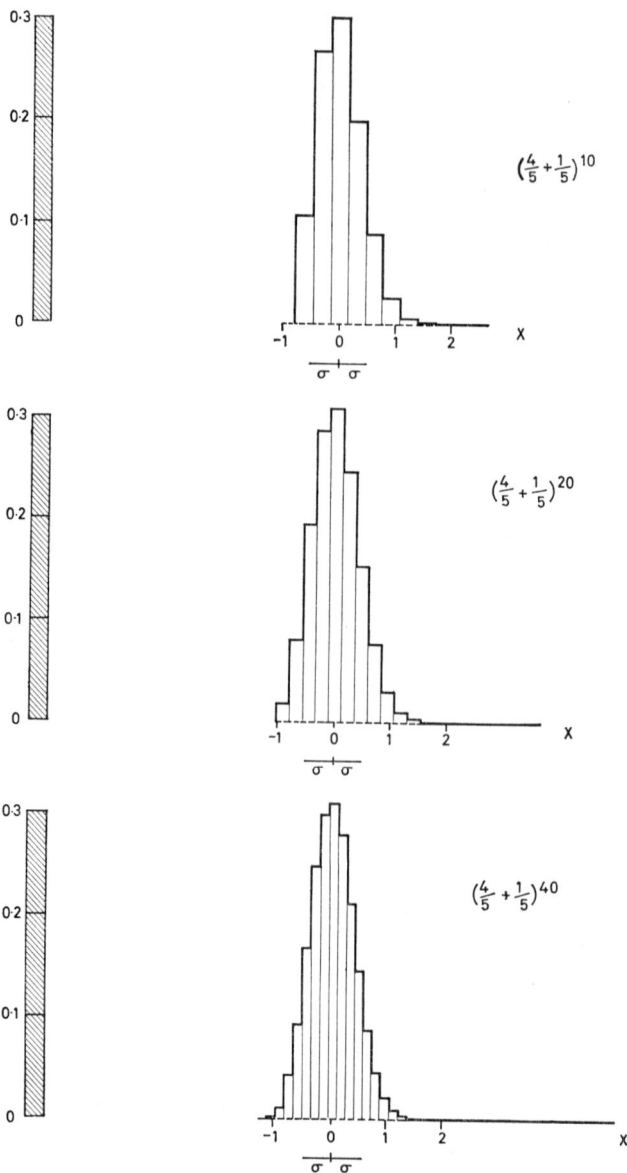

Fig. 8.2

TABLE OF PARTIAL AREAS UNDER THE NORMAL CURVE

$$\text{Values of } A(X) = \frac{1}{\sigma\sqrt{(2\pi)}} \int_0^x e^{-x^2/2\sigma^2}\, dx$$

X/σ	0	1	2	3	4	5	6	7	8	9
0·0	·0000	·0040	·0080	·0120	·0159	·0199	·0239	·0279	·0319	·0359
0·1	·0398	·0438	·0478	·0517	·0557	·0596	·0636	·0675	·0714	·0753
0·2	·0793	·0832	·0871	·0910	·0948	·0987	·1026	·1064	·1103	·1141
0·3	·1179	·1217	·1255	·1293	·1331	·1368	·1406	·1443	·1480	·1517
0·4	·1554	·1591	·1628	·1664	·1700	·1736	·1772	·1808	·1844	·1879
0·5	·1915	·1950	·1985	·2019	·2054	·2088	·2123	·2157	·2190	·2224
0·6	·2257	·2291	·2324	·2357	·2389	·2422	·2454	·2486	·2517	·2549
0·7	·2580	·2611	·2642	·2673	·2704	·2734	·2764	·2794	·2823	·2852
0·8	·2881	·2910	·2939	·2967	·2995	·3023	·3051	·3078	·3106	·3133
0·9	·3159	·3186	·3212	·3238	·3264	·3289	·3315	·3340	·3365	·3389
1·0	·3413	·3438	·3461	·3485	·3508	·3531	·3554	·3577	·3599	·3621
1·1	·3643	·3665	·3686	·3708	·3729	·3749	·3770	·3790	·3810	·3830
1·2	·3849	·3869	·3888	·3907	·3925	·3944	·3962	·3980	·3997	·4015
1·3	·4032	·4049	·4066	·4082	·4099	·4115	·4131	·4147	·4162	·4177
1·4	·4192	·4207	·4222	·4236	·4251	·4265	·4279	·4292	·4306	·4319
1·5	·4332	·4345	·4357	·4370	·4382	·4394	·4406	·4418	·4430	·4441
1·6	·4452	·4463	·4474	·4484	·4495	·4505	·4515	·4525	·4535	·4545
1·7	·4554	·4564	·4573	·4582	·4591	·4599	·4608	·4616	·4625	·4633
1·8	·4641	·4649	·4656	·4664	·4671	·4678	·4686	·4693	·4699	·4706
1·9	·4713	·4719	·4726	·4732	·4738	·4744	·4750	·4756	·4762	·4767
2·0	·4772	·4778	·4783	·4788	·4793	·4798	·4803	·4808	·4812	·4817
2·1	·4821	·4826	·4830	·4834	·4838	·4842	·4846	·4850	·4854	·4857
2·2	·4861	·4864	·4868	·4871	·4875	·4878	·4881	·4884	·4887	·4890
2·3	·4893	·4896	·4898	·4901	·4904	·4906	·4909	·4911	·4913	·4916
2·4	·4918	·4920	·4922	·4925	·4927	·4929	·4931	·4932	·4934	·4936
2·5	·4938	·4940	·4941	·4943	·4945	·4946	·4948	·4949	·4951	·4952
2·6	·4953	·4955	·4956	·4957	·4959	·4960	·4961	·4962	·4963	·4964
2·7	·4965	·4966	·4967	·4968	·4969	·4970	·4971	·4972	·4973	·4974
2·8	·4974	·4975	·4976	·4977	·4977	·4978	·4979	·4980	·4980	·4981
2·9	·4981	·4982	·4982	·4983	·4984	·4984	·4985	·4985	·4986	·4986
3·0	·4987	·4987	·4987	·4988	·4988	·4989	·4989	·4989	·4990	·4990
3·1	·4990	·4991	·4991	·4991	·4992	·4992	·4992	·4992	·4993	·4993

a frequency distribution the range $m - 2s$ to $m + 2s$ usually covers about 95 per cent of the values of the variable. Similarly, the table shows that the probability of a deviation either way of more than 3σ is $2(0.5 - 0.4987)$, or 0.0026, i.e. 1 in 400 approx. In so far as frequency curves resemble the normal curve (which is often the case), these figures give a more precise meaning to the rough rule of Chapter 2.

Total area under the normal curve

Since the area under the curve represents the probability of all possible values, the total area should be one unit. This is confirmed by the derivation of the curve from the terms of $(q + p)^n$, as shown at the end of the chapter. It follows, putting $\sigma = 1$, that the area under $y = e^{-\frac{1}{2}x^2}$ is $\sqrt{(2\pi)}$; or, putting $\sigma^2 = \frac{1}{2}$, that the area under $y = e^{-x^2}$ is $\sqrt{\pi}$.

* Readers familiar with double integrals may like to see this proved directly, as follows:

If
$$I = \int_{-\infty}^{\infty} e^{-x^2} \, dx = \int_{-\infty}^{\infty} e^{-y^2} \, dy,$$

$$I^2 = 4 \int_{0}^{\infty} e^{-x^2} \, dx \int_{0}^{\infty} e^{-y^2} \, dy,$$

$$= 4 \int_{0}^{\infty} \int_{0}^{\infty} e^{-(x^2+y^2)} \, dx \, dy.$$

Changing to polar coordinates, this is $4 \int_{0}^{\pi/2} \int_{0}^{\infty} e^{-r^2} r \, dr \, d\theta,$

$$= 2 \int_{0}^{\pi/2} d\theta,$$

$$= \pi.$$

$$\therefore I = \int_{-\infty}^{\infty} e^{-x^2} \, dx = \sqrt{\pi}.$$

Use of the normal curve

The most direct use of the normal curve is as an approximation to the binomial distribution when n is large. We can use the table of areas to answer such questions as the following:

Example 8.1

If a coin is spun 500 times, what is the probability of getting more than 265 'heads'?

Expected number of 'heads' = 250.
Excess = 15. S.D. = $\sqrt{(500 \times \frac{1}{2} \times \frac{1}{2})}$ = 11·18 approx.
$\therefore$ excess = 1·342 × S.D.

From the table, the area beyond the ordinate $X/\sigma = 1·342$ is

0·5 − 0·4102, or 0·0898. The probability is, therefore, 0·09 approx., or roughly 1 in 11.

Note. If we ask 'What is the probability of a deviation of more than 15 either way?' the answer is, of course, twice as much.

Example 8.2

If an integer is chosen at random the chance of its being divisible by 5 is $\frac{1}{5}$. *If 20 telephone numbers are picked at random from the directory, what is the probability that at least 6 of them should be divisible by 5?*

Here $n = 20$ and, with such a small value, it is hardly to be expected that the normal curve will give a close approximation to the desired result. Nevertheless, something can be done. In Fig. 8.3 the rectangles

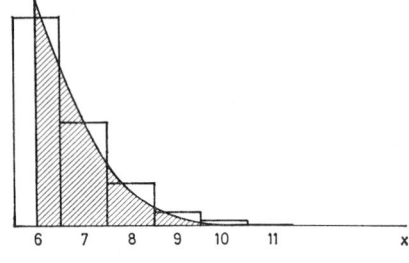

Fig. 8.3

represent part of the binomial distribution for $n = 20$ and $p = \frac{1}{5}$, with a normal curve fitted. The true probability of 6 or more successes is represented by the total area of the rectangles shown. To estimate this by means of the normal curve we note first that the deviation from the expected value 4 is 2. The estimate obtained by using the table of areas, with $x/\sigma = 2/\sqrt{(npq)}$, is represented by the shaded area under the curve. It will be seen that the major part of the error is represented by the area of the half-rectangle on the extreme left of the diagram, and that this is an appreciable fraction of the area under consideration. A better estimate can be obtained by taking the area under the curve from $x = 5\frac{1}{2}$ onwards.

Using this method, we have $X/\sigma = 1\cdot5/\sqrt{(20 \times \frac{4}{5} \times \frac{1}{5})} = 0\cdot838$ approx. From the table of areas, the probability of 6 or more successes is then $0\cdot5 − 0\cdot299$, i.e. $0\cdot201$, or $0\cdot2$ approx.

(The true result, worked from the terms of the binomial expansion, is 0·195.)

Frequency distributions and the normal curve

Many frequency distributions are unimodal and symmetrical and approximate in form to the normal curve. A good example is found in the distribution of height among a large group of men. Intelligence

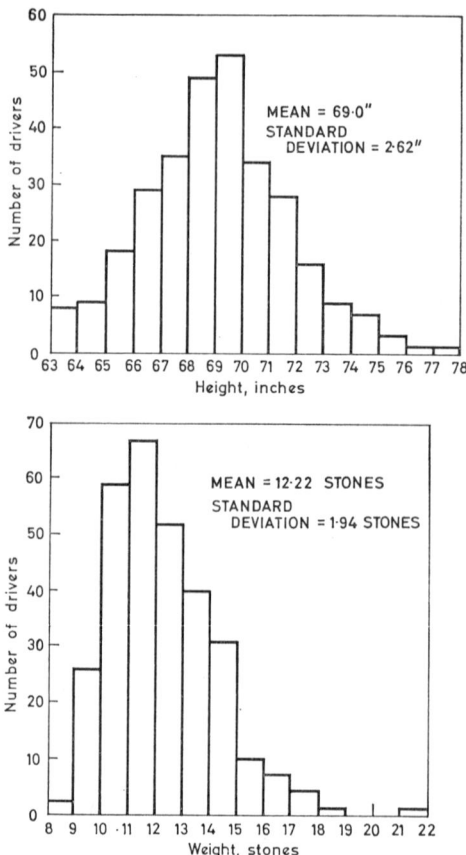

Fig. 8.4. Heights and weights of 300 lorry drivers (Reproduced by courtesy of the Director of the Motor Industries Research Association)

quotients, examination marks, and the variations in size of mass-produced articles frequently follow more or less closely the same pattern. It is natural to enquire why this should be so, but it is hardly possible to do more than hazard a guess at the reason. Indeed, the very fact that some distributions are skew or differ in other ways from the so-called 'normal' pattern indicates that there is no compelling reason why all distributions should tend towards this type.

For example, if heights follow the normal distribution, the same cannot apply to weights, since they are by no means proportional to heights but more nearly to their cubes. This is illustrated in Fig. 8.4.

Bearing in mind the derivation of the normal curve from the binomial distribution, it would seem likely to occur when the variations are the combined result of large numbers of equal small variations, each of which may be positive or negative. A man's height might, conceivably, be due to the interaction of a large number of factors, e.g. his daily growth during a period of years, each one tending to produce a slightly greater or less eventual height.

It is best to regard it as an empirical fact that a good many distributions are sufficiently near to the 'normal' for the normal curve to be used as a model for them.

Errors of observation

Gauss (1777–1855) put forward a 'proof' that observational errors tend to follow the normal law. He assumed, first, that the probability of any particular observed value lying in a small interval would be a definition function $\phi(\Delta)$ of its deviation Δ from the true value. The probability of a set of observed values could thus be expressed in terms of ϕ and of the supposed position of the true value. He assumed, further, that the likeliest position of the true value would be such as to make this probability a maximum; and, moreover, that it would coincide with the mean of the observations. In this way he ingeniously derived the equation of the normal curve (and it has been called 'the Gaussian curve' by many writers since then, although it had been discovered by de Moivre about a hundred years earlier). But, as pointed out by Bertrand (*Calcul des Probabilités*, p. 175), the method cannot be valid. If it is assumed, as in Gauss's method, that the most probable true value is the mean of the observed values, why should not the same apply to the squares or the logarithms of those values? Why should not the most probable value of the square of the true value be the mean of the squares of the observations? Thus, the idea leads to contradictions.

Nevertheless, it is an empirical fact that observational errors do tend to follow the 'Gaussian law', and the normal curve may be usefully employed in dealing with them.

The mean of a sample

It has already been shown (p. 60) that if a sample of n members is taken from a population whose standard deviation is σ, the mean of the sample is distributed with S.D. $\sigma/\sqrt{n}$. It is, furthermore, found that for a large sample, the mean is distributed approximately according to the normal curve, even when the population from which the sample is drawn is by no means normal. This is illustrated in Exercise 12, Nos 1, 2, 3 (p. 123).

Example 8.3

Holes are drilled with mean diameter 1·5 cm, and S.D. 0·03 cm, and bolts are made with mean diameter 1·4 cm. and S.D. 0·02 cm. Assuming normal distribution, find the expected proportion of bolts that will be too large for holes picked at random.

The mean clearance will be 0·1 cm with S.D. $\sqrt{(0·03^2 + 0·02^2)}$, i.e. 0·036 cm (*see* Chapter 5, p. 58) For a bolt not to fit, the deviation from the mean would have to be at least 0·1 cm, which is 2·78 times the S.D. From the table of areas, the chance of a deviation greater than this is 0·5 − 0·4973, or 0·0027. It might, therefore, be expected that, in random choice, about 0·27 per cent of the bolts would be too large for their holes.

Example 8.4

If the mean height of a man is 5 ft 9 in, with a standard deviation 2·5 in, find the probability that the average height of 20 men chosen at random should be more than 5 ft 10 in.

$$\text{S.D. of the mean} = \frac{2·5}{\sqrt{20}} = 0·559 \text{ in.}$$

$$\therefore \text{ a deviation of 1 inch} = \frac{1}{0·559} \times \text{S.D.,}$$

$$= 1·789 \times \text{S.D.}$$

From the table of areas, the probability of a deviation greater than this in the positive direction is 0·5 − 0·463, or 0·037, or 1 in 27 approx.

EXERCISE 10

(*Assume normal distribution in all these examples.*)

1. Electric lamps of a certain type burn for 1800 hours on the average, the standard deviation being 42 hours. What percentage are likely to

The normal distribution

fail before completing 1700 hours? For how long would you expect the best 10 per cent of the lamps to burn?

2. In a large examination, the average mark of candidates is 54·4 per cent, with a S.D. of 15 per cent. What percentage of candidates are likely to have marks of 40 per cent or over? What percentage are likely to have 80 per cent or over?

3. If the I.Q.s (intelligence quotients) of the schoolchildren in a certain country give a mean of 105, with S.D. 12·8, how many in every thousand would you expect to find with I.Q.s below 100? How high would be the I.Q.s of the top 25 per cent?

4. Use the data of Exercise 5, No. 4 (p. 63) to find the percentage of drivers in each group likely to have driven at an average of 32·5 mile/hr or more.

    (*Note.* The first group were drivers who had been convicted for careless driving during the previous 3 years; the second group were selected at random. Actual numbers in the groups tested were 10 and 7.)

5. If the mean height of a man is 5 ft 9 in, with a S.D. of $2\frac{1}{2}$ in, what percentage are over 6 ft? What is the probability that the mean height of 25 men would be between 5 ft 8 in and 5 ft 10 in?

6. Castings are produced in quantity with a mean weight of 14·3 kg and S.D. 0·2 kg. What is the chance of a particular one being over 14·6 kg? What is the chance that the mean weight of 10 of them should differ from 14·3 by more than 0·1 kg?

7. Electric lamps burn for a mean time of 1500 hours with a S.D. of 60 hours.

    (i) What is the probability of a lamp failing before it has burned for 1400 hours?

    (ii) What percentage of lamps would you expect to last more than 1650 hours?

    (iii) If the average life of a sample of 20 lamps is 1540 hours, would you regard the result as remarkably good?

8. Bolts are manufactured to fit into holes in steel plates. The mean diameter of the bolts is 3·20 cm, with S.D. 0·03 cm; that of the holes is 3·31 cm, with a S.D. of 0·04 cm.

    (i) Find the proportion of bolts with diameter greater than 3·25 cm.

    (ii) Find the proportion of holes with diameter less than 3·25 cm.

    (iii) If the bolts and holes are selected at random, show that the proportion of bolts that will not fit the holes is 1·39 per cent.

    (iv) Find the chance that a batch of 100 bolts will all fit their holes.

9. If the average height of a man is 5 ft 9 in, with S.D. 2·5 in, and the average height of a woman is 5 ft 4 in, with the same S.D., find the probability that, of a man and a woman, both picked at random, the woman should be the taller.

10. In a mechanism, a plunger moves inside a cylinder, and the difference between the diameters must be at least 0·01 cm. The diameters are

distributed about a mean of 2·025 cm, with S.D. 0·002 cm; the diameters of the plungers are distributed about a mean of 2·010 cm, with S.D. 0·003 cm.
  (i) Find the proportion of cylinders that will have inadequate clearance for plungers of diameter 2·01 cm.
  (ii) Find the proportion of plungers that will have inadequate clearance in cylinders of diameter 2·025 cm.
  (iii) If plungers and cylinders are assembled at random, find the proportion of cases in which the clearance is insufficient.
  (iv) If a set of 10 plungers and cylinders is assembled in random pairs, find the chance of the clearance being sufficient in all cases.
11. In an examination, there are 4000 candidates. The marks may be taken as a continuous and normal distribution with mean 47·5 and standard deviation 12·5.
  (i) Estimate the number of candidates who obtain marks between 36 and 44.
  (ii) Estimate the mark that will be exceeded by 10 per cent of the candidates.
  (iii) If a particular examiner marks 100 scripts taken at random and awards marks whose mean is 49·5, is there evidence that his marking is more lenient than that of the other examiners?
12. Steel bolts of circular cross-section are being manufactured to a specification which requires that their lengths be between 8·45 and 8·65 cm, and their diameters between 1·55 and 1·60 cm. A machine produces these bolts so that their lengths have a mean of 8·54 cm, with S.D. 0·05 cm, and their diameters are independently distributed about a mean of 1·57 cm, with S.D. 0·01 cm.
  (i) Find the percentage of bolts that will not be within the specified limits for (*a*) length, (*b*) diameter.
  (ii) Find the percentage of bolts that will not meet the specifications.
  (iii) Find the probability that in a sample of five bolts taken at random four should meet the requirements and one should fail.
13. Steel shafts are manufactured with diameters that must lie between 1·75 and 1·85 cm. 2 per cent have to be rejected as being too large and another 2 per cent as being too small. Estimate the standard deviation and find what percentage would have to be rejected if the tolerance limits were 1·77 cm and 1·83 cm.
14. In an examination, the pass mark is 40 per cent and the distinction mark is 75 per cent. If 70·5 per cent of the candidates pass and 10 per cent gain distinction, estimate the mean mark and the standard deviation. What percentage of candidates would you expect to find with marks of more than 60 per cent?
15. Transistors bought in batches of 1000 show an average 'current gain' (β) of 100, with S.D. 15·2. For a special purpose a transistor with $\beta \geq 150$ is required. What is the probability of finding one or more in a batch of 1000?

Fitting a normal curve

Suppose that, for a given frequency diagram, it is desired to draw a normal curve to fit as closely as possible to the recorded distribution. The mean of the distribution and its standard deviation are first evaluated, and these determine the two parameters of the normal curve, namely the position of the centre and the standard deviation σ. The curve can then be plotted from a table of ordinates. For this purpose a very short table is sufficient, as follows:

Values of $\frac{x}{\sigma}$ (measured from centre)	Values of $\frac{1}{\sqrt{(2\pi)}} e^{-x^2/(2\sigma^2)}$
0	0·3989
0·25	0·3862
0·5	0·3521
1·0	0·2420
1·5	0·1295
2·0	0·0540
2·5	0·0175
3·0	0·0044

The scale for x must be chosen with its zero at the mean. The values of the ordinates must be divided by σ (as measured on the x-scale) and then multiplied by N, the total frequency, to ensure that the total area under the curve will represent N.

Example 8.5

The disintegrations of some radioactive silver occurring during 15-second periods were counted by means of a Geiger–Muller tube and a Dekatron counter, with the following results:

No. of disintegrations	Frequency
210–219	5
220–229	10
230–239	22
240–249	24
250–259	28
260–269	20
270–279	8
280–289	3
	120

104 Statistics: the how and the why

From these figures the mean is found to be 248·4, and the standard deviation 16·25 (or 1·625, using the group interval of 10 as unit of x). The ordinates given in the table are, therefore, multiplied by 120/1·625. This gives:

x/σ	Ordinate to be plotted
0	29·5
0·25	28·6
0·5	26·1
1·0	17·9
1·5	9·6
2·0	4·0
2·5	1·3
3·0	0·3

The result is shown in Fig. 8.5.

It may often, however, be of more interest to work out expected frequencies, on the hypothesis of a normal distribution, for comparison with the observed frequencies. It is then necessary to tabulate the values of x/σ corresponding to the boundaries of the groups; to find the areas under the normal curve between these values; and to multiply them by the total frequency.

Boundary	x (measured from mean)	x/σ (Divide x by 1·625)	Area (from table)	Diff.	Expected freq. (Mult. diff. by 120)	Observed freq.
209·5	−3·89	−2·394	−0·4917			
219·5	−2·89	−1·778	−0·4623	0·0294	3·5	5
229·5	−1·89	−1·163	−0·3776	0·0847	10·2	10
239·5	−0·89	−0·548	−0·2081	0·1695	20·3	22
249·5	0·11	0·068	0·0271	0·2352	28·2	24
259·5	1·11	0·683	0·2527	0·2256	27·1	28
269·5	2·11	1·298	0·4029	0·1502	18·0	20
279·5	3·11	1·914	0·4722	0·0693	8·3	8
289·5	4·11	2·529	0·4943	0·0221	2·7	3

Normal probability graph paper

This is paper for drawing cumulative frequency curves, with the scale so arranged that the curve for a normal distribution will be a straight line. Figure 8.6a shows such a curve plotted on ordinary paper, and Fig. 8.6b the curve for the same distribution plotted on 'normal probability' paper.

The method of construction of this paper is as follows: If the area under the normal curve is divided by ordinates into 100 equal parts

The normal distribution   105

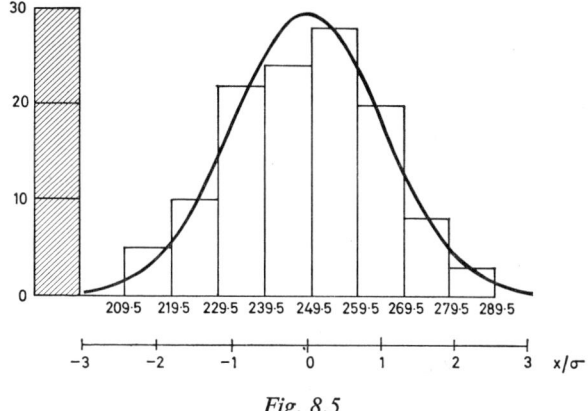

Fig. 8.5

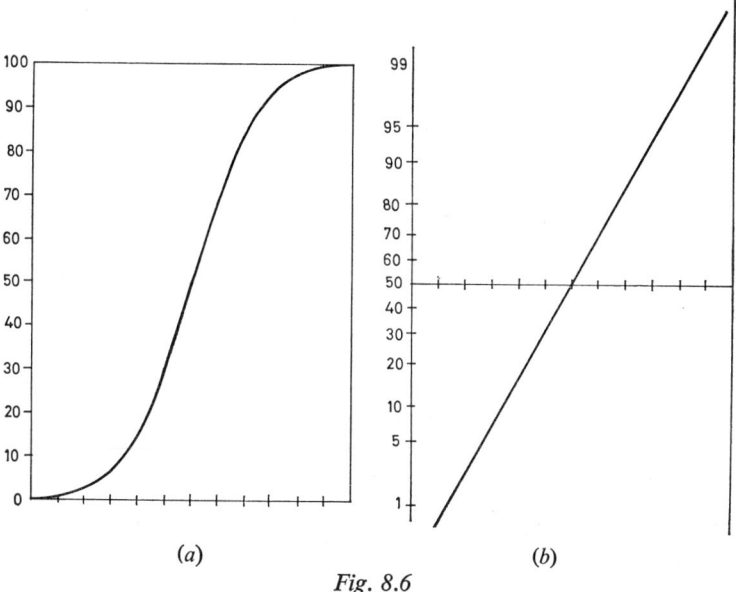

Fig. 8.6

(Fig. 8.7) the scale so obtained might be described as one of cumulative probability; that is to say that the ordinate numbered 60, for example, coincides with the value of X/σ such that there is a 60 per cent chance of the value not being exceeded.

8

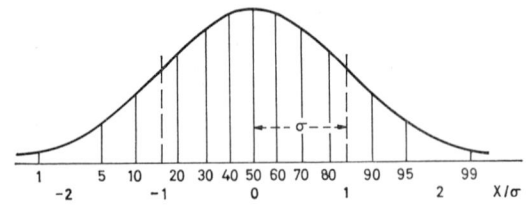

Fig. 8.7

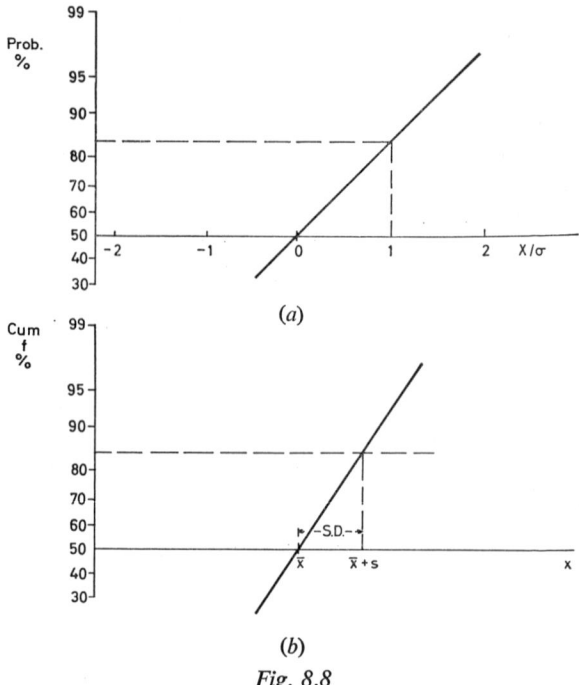

Fig. 8.8

Normal probability paper (sometimes called 'arithmetical proability paper') is made with this scale on the 'vertical' axis and a uniform scale on the 'horizontal' axis. The effect is that, if the cumulative probabilities of the normal distribution are plotted on this 'vertical' scale against the corresponding values of X/σ on the 'horizontal' scale (Fig. 8.8a), the points lie on a straight line.

Suppose now that the 'horizontal' axis is re-labelled to accommodate, on any convenient uniform scale, the values of a variable x that is normally distributed, and that the cumulative frequencies, expressed as percentages, are plotted (Fig. 8.8b). The resulting points will again lie on a straight line, cutting the 50 per cent line at the mean value $\bar{x}$ of x, its gradient depending on the standard deviation and on the scale chosen. The values of $(\bar{x} \pm s)$ correspond to cumulative frequencies of $50 \pm 34\cdot13$ per cent, and from this the standard deviation of the distribution can be estimated.

Using the figures of Example 8.5 (p. 103) we first tabulate the cumulative frequencies and reduce them to percentages:

Group	f	Cum. f.	Per cent
210–219	5	5	4·2
220–229	10	15	12·5
230–239	22	37	30·8
240–249	24	61	50·8
250–259	28	89	74·2
260–269	20	109	90·8
270–279	8	117	97·5
280–289	3	120	100·0
	120		

The percentage cumulative frequencies are plotted against the upper boundaries of the groups, using any convenient scale on the 'horizontal' axis (Fig. 8.9). The mean is seen to be approximately 248·5, and the standard deviation $\frac{1}{2}(264\cdot5 - 232\cdot0)$, i.e. 16·25, in close agreement with the calculated values (p. 104).

EXERCISE 11

1. Copy the histogram of the heights of lorry drivers (Fig. 8.4, p. 98) and draw a normal curve fitting it as closely as possible. Calculate, also, the expected frequencies, on the hypothesis of a normal distribution, and compare them with the observed frequencies.
2. The breaking stresses in ton/in^2 of 200 steel bars were as follows:

Breaking stress	Frequency
29·5–29·7	12
29·7–29·9	27
29·9–30·1	60
30·1–30·3	58
30·3–30·5	30
30·5–30·7	13

Draw a histogram and fit a normal curve. Calculate the expected frequencies for a normal distribution.

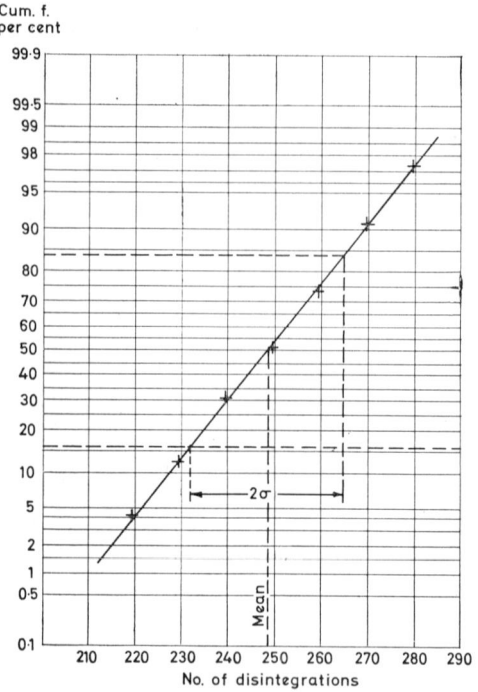

Fig. 8.9. Disintegration of radioactive silver

3. Use normal probability paper to test whether the marks of the examination candidates in Exercise 1, p. 22, No. 13, were normally distributed. Find the expected frequencies for a normal distribution. (This may be done approximately with the aid of the graph, or the values may be calculated, using the calculated mean 45·0 and S.D. 11·4.)

4. The following table shows the intelligence quotients of 228 boys:

I.Q.	Frequency	I.Q.	Frequency
85–89	12	110–114	38
90–94	11	115–119	36
95–99	20	120–124	21
100–104	32	125–129	10
105–109	39	130–134	7
		135–139	2

Use normal probability paper to test whether the distribution is approximately normal, and estimate the mean and standard deviation.

The normal distribution

5. Plot points on probability paper for distributions known to be skew, e.g. the flying bombs (Exercise 9, p. 89, No. 7) (positively skew) and the number of references per page (Chapter 2, p. 9, Example 2.1) (negatively skew).

6. Make a histogram for the lengths of the broad bean seeds given in Exercise 1, p. 21, No. 11. Assuming that the mean is 2·70 cm, and the S.D. 0·22 cm, draw a normal curve to fit as closely as possible. Calculate expected frequencies for a normal distribution.

7. In an experiment, 700 lead shot were run down a sloping board in which pins had been inserted at regular intervals. On hitting a pin, a shot was deflected one way or the other, so it might be anticipated that the distribution of the shot would be approximately normal. The shot were collected in pens uniformly spaced along the bottom of the board and were counted, with the following results:

Pen no.	No. of shot	Pen no.	No. of shot
3	2	14	87
4	8	15	58
5	7	16	47
6	6	17	52
7	30	18	30
8	29	19	26
9	43	20	15
10	54	21	16
11	35	22	7
12	69	23	7
13	72		

Use normal probability paper to examine these figures.

*Equation of the normal curve

Suppose that in Fig. 8.2 a curve is drawn through the mid-points of the tops of the rectangles. Figure 8.10 shows the diagrams for $(\frac{4}{5} + \frac{1}{5})^{20}$ and $(\frac{4}{5} + \frac{1}{5})^{40}$, with curves so drawn. It will be remembered that, as n increases, the scale on the 'horizontal' axis is reduced in the ratio $1/\sqrt{n}$, and the standard deviation $\sqrt{(npq)}$ is thus represented by a constant length $\sqrt{(pq)}$. Taking the base-line as the axis of X, with the origin at the mean number of successes, np, the value of X for r successes is given by

$$X = \frac{r - np}{\sqrt{n}}, \quad \text{or} \quad r = np + X\sqrt{n} \quad (\text{see Fig. 8.10}) \qquad (1)$$

The standard deviation is then $\sqrt{(pq)}$ ($=0.4$ in this particular case), measured on the X-scale. The term $_nC_r q^{n-r} p^r$ (giving the probability of r successes out of n) is represented by the area of the rectangle at the point X, as defined above.

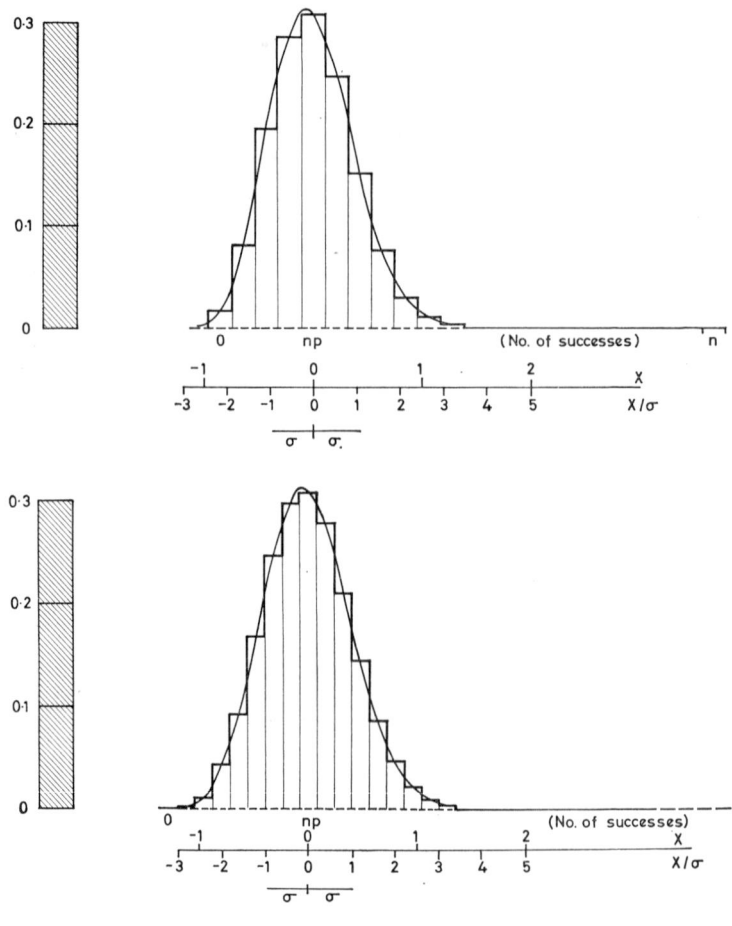

Fig. 8.10

If the ordinates of two consecutive plotted points are Y_{r-1} and Y_r the average gradient between them is given by

$$\frac{\delta Y}{\delta X} = \frac{Y_r - Y_{r-1}}{1/\sqrt{n}},$$

and in the limit, as n increases indefinitely,

$$\frac{1}{Y}\frac{dY}{dX} = \operatorname*{Lt}_{n \to \infty} \frac{Y_r - Y_{r-1}}{Y_{r-1}} \cdot \sqrt{n} = \operatorname*{Lt}_{n \to \infty} \left(\frac{Y_r}{Y_{r-1}} - 1\right) \cdot \sqrt{n}.$$

But $\left(\dfrac{Y_r}{Y_{r-1}} - 1\right).\sqrt{n} = \left(\dfrac{n-r+1}{r}.\dfrac{p}{q} - 1\right).\sqrt{n},$

$$= \dfrac{np - r + p}{rq}.\sqrt{n},$$

since $p + q = 1$.
Using (1), this is equal to

$$\dfrac{-X\sqrt{n} + p}{npq + X\sqrt{n}.q}.\sqrt{n},$$

and the limit of this, as $n \to \infty$, is

$$\dfrac{-X}{pq}, \quad \text{or} \quad \dfrac{-X}{\sigma^2}.$$

$$\therefore \dfrac{1}{Y}\dfrac{dY}{dX} = -\dfrac{X}{\sigma^2}.$$

Integrating with respect to X,

$$Y = Ae^{-X^2/(2\sigma^2)}.$$

The constant A can be determined from the fact that the whole area under the curve is 1. Putting $X = \sqrt{2}.\sigma u$.

$$\int_{-\infty}^{\infty} Y\,dX = A\int_{-\infty}^{\infty} e^{-u^2}.\sqrt{2}.\sigma\,du, = A\sqrt{(2\pi\sigma^2)}. \quad \text{(see p. 96)}$$

Hence

$$A = \dfrac{1}{\sqrt{(2\pi\sigma^2)}} \quad \text{and} \quad Y = \dfrac{1}{\sqrt{(2\pi\sigma^2)}}e^{-X^2/(2\sigma^2)}.$$

*Distribution of the sum of two normal variates

It has already been proved (p. 57) that the mean value of the sum of two independent variates is equal to the sum of their means. We now consider deviations from the mean.

Let x, y be independent variates with normal probability distributions $\phi(x), \psi(y)$ respectively, where

$$\phi(x) = \dfrac{1}{\sqrt{(2\pi\sigma_x^2)}} e^{-\frac{1}{2}x^2/\sigma_x^2} \quad \text{and} \quad \psi(y) = \dfrac{1}{\sqrt{(2\pi\sigma_y^2)}} e^{-\frac{1}{2}y^2/\sigma_y^2}.$$

Suppose that x and y are represented on rectangular axes on a horizontal plane and that z is measured vertically. The compound probability that x should lie in the interval $[x, x + dx]$ and y in the interval [y, y + dy] is $\phi(x)\psi(y)\,dx\,dy$, and is represented by an element of volume between the surface

$$z = \phi(x)\psi(y), \quad = \frac{1}{2\pi\sigma_x\sigma_y} e^{-\frac{1}{2}\left(\frac{x^2}{\sigma_x^2} + \frac{y^2}{\sigma_y^2}\right)}$$

and the rectangle $dx\,dy$ in the xy plane.

This surface takes the form of an oval-shaped hillock rising out of the xy plane, with the z-axis at its centre. Its contours are elliptical, with semi-axes proportional to σ_x, σ_y (Fig. 8.11). The probability that

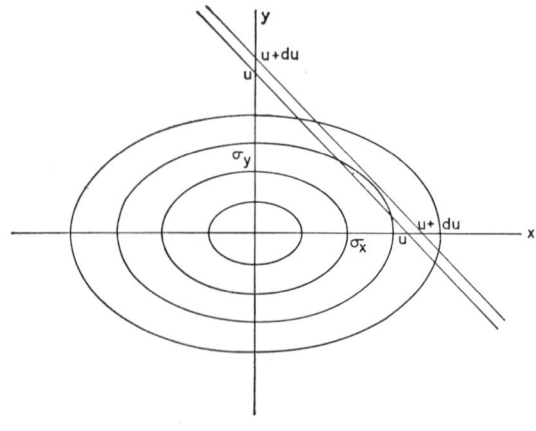

Fig. 8.11

$(x + y)$ should lie in the fixed interval $[u, u + du]$ is represented by the volume in the section of the hillock lying between the planes

$$x + y = u \quad \text{and} \quad x + y = u + du.$$

To measure this volume, we make the transformation

$$x = \sigma_x X, \quad y = \sigma_y Y.$$

Thus, z becomes

$$\frac{1}{2\pi\sigma_x\sigma_y} e^{-\frac{1}{2}(X^2 + Y^2)}$$

and the surface is now a surface of revolution about the z-axis (Fig. 8.12). A section of the volume now depends solely on its distance r from the z-axis and its thickness dr.

The line $x + y = u$, marking the 'inner' boundary of the section,

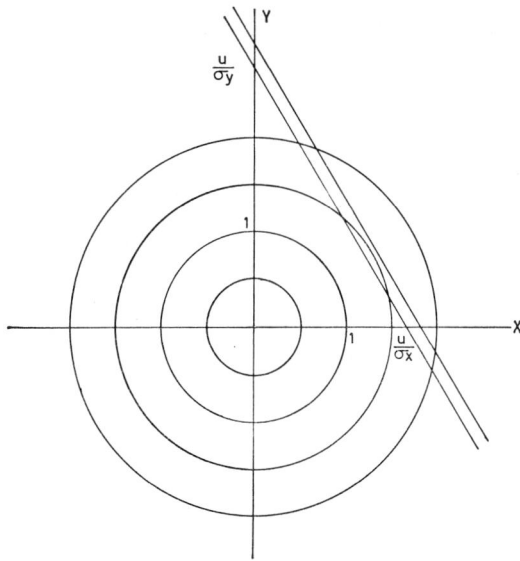

Fig. 8.12

is now $\sigma_x X + \sigma_y Y = u$ (Fig. 8.12), and its distance from the origin, r, is

$$\frac{u}{\sqrt{(\sigma_x^2 + \sigma_y^2)}}, \quad \text{with} \quad dr = \frac{du}{\sqrt{(\sigma_x^2 + \sigma_y^2)}}.$$

The volume is the same as if the section were taken parallel to the Y-axis at distance r from the origin, with thickness dr. But the area of such a section is

$$\frac{1}{2\pi\sigma_x\sigma_y} e^{-\frac{1}{2}r^2} \int_{-\infty}^{\infty} e^{-\frac{1}{2}Y^2} \, dY, \quad \text{i.e.} \quad \frac{1}{\sqrt{(2\pi)}} \frac{1}{\sigma_x\sigma_y} e^{-\frac{1}{2}r^2},$$

and the volume is

$$\frac{1}{\sqrt{(2\pi)}} \frac{1}{\sigma_x\sigma_y} e^{-\frac{1}{2}r^2} \, dr,$$

i.e.

$$\frac{1}{\sqrt{(2\pi)}} \frac{1}{\sigma_x\sigma_y} e^{-\frac{1}{2}\frac{u^2}{\sigma_x^2+\sigma_y^2}} \cdot \frac{du}{\sqrt{(\sigma_x^2 + \sigma_y^2)}}.$$

The corresponding volume on the original diagram is $\sigma_x\sigma_y$ times as much, i.e.

$$\frac{1}{\sqrt{(2\pi)}\sqrt{(\sigma_x^2 + \sigma_y^2)}} e^{-\frac{1}{2}\frac{u^2}{\sigma_x^2+\sigma_y^2}} \, du.$$

Thus, the distribution of u is normal, with variance $(\sigma_x^2 + \sigma_y^2)$.

The theorem may be extended to more than two variables. Thus, *the sum of any number of normal variates is distributed normally about a mean equal to the sum of their means, with a variance equal to the sum of their variances.*

It will also be seen that the difference of two normal variates is distributed normally about a mean equal to the difference of their means, with a variance equal to the sum of their variances.

9
Sampling

There are two main problems in sampling. First, if the distribution of a 'population' is known, what results may be expected when we take samples from it? Secondly, there is the inverse problem: if samples are taken and the results recorded, what can be inferred about the distribution of the population? The second problem is, in fact, treated by means of the first, that is to say, by setting up a hypothetical population and considering the probability of obtaining the observed results by sampling from it.

Terminology and notation

A *population* is the complete set of N values of a variable x from which a *sample*, $x_1, x_2, x_3, \ldots, x_n$, is chosen, n being the number of members in the sample. To distinguish the characteristics of the sample from those of the population, the following notation will be used:

	Population	Sample
Number of members	N	n
Mean	μ	$\bar{x}$
Standard deviation	σ	s

Such numbers as μ and σ are called the *parameters* of the population, while $\bar{x}$ and s and other measures descriptive of the sample are called the *statistics* of the sample.

Populations

The population may be actual or hypothetical. An actual population might, for example, be the lengths of mass-produced components turned out by a factory in the course of a day. A hypothetical population might be the lengths of the components that would be turned out by an idealized factory whose machines worked in some particular way, e.g. if the mean length of the components were 10 cm, and the standard deviation 0·05 cm, according to standards laid down by the works manager.

To measure an actual population completely is often impossible, either because it would take too long, or because testing destroys the product (as with explosives or tinned foods), or because the whole population is not available (as, for example, with the population consisting of the weights of all Alsatian dogs, or all possible throws of a die). For such reasons samples are taken, and the statistics of the samples may be used in an attempt to estimate the parameters of the population.

Random samples

A *random sample* is one in which every member of the population has the same chance of being chosen. The practical problem of picking a random sample is often difficult, and various techniques are employed. Such primitive methods as drawing names out of a hat or shuffling cards or tossing up are of somewhat limited application. A better method is to use a table of random numbers, such as the following:

58	91	05	97	46	96	86	86	90	64	77	84
40	85	97	85	95	07	41	00	83	93	41	56
12	78	87	44	42	65	82	53	46	78	57	87
90	21	67	51	83	51	00	78	38	65	24	91
25	02	71	93	90	31	18	14	73	36	03	68
04	32	93	96	19	56	20	92	39	88	98	06
92	41	93	34	33	13	07	37	82	22	21	97
34	01	20	79	03	57	32	66	04	79	55	73
91	01	71	40	84	35	42	64	89	74	18	98
28	65	56	27	81	05	17	28	22	79	96	33
93	55	64	77	53	82	30	59	36	13	35	20
73	65	71	30	03	38	21	38	01	64	52	58

This table is hardly large enough for practical purposes, but will serve to illustrate the method. Books of random numbers are published (unique among books, in that they consist entirely of figures arranged in complete disorder, without meaning or significance of any kind). The tables of winning premium bonds, published from time to time in the press, could be used for the same purpose.

If it is desired to pick a random sample from, say, 500 objects, it is first necessary to number them. The numbers of the objects to be picked for the sample can then be taken from the table in any

Sampling

systematic manner, e.g. the digits could be taken in threes from the rows or the columns, any combination of digits giving a number greater than 500 being rejected.

Sampling of a variable

It has already been shown (p. 59) that the mean $\bar{x}$ of a random sample of n members has a probability distribution with expected value μ and standard deviation $\sigma/\sqrt{n}$. This applies to all types of population and to all sizes of sample, large or small.

The difference ($\bar{x} - \mu$) between the mean of the sample and that of the population is called the *error of the mean*, and its standard deviation is called the *standard error of the mean*. We have used the rough rule that the error is not likely to be more than twice the standard error and will hardly ever exceed three times the standard error; but to draw more exact conclusions we must know more about the distribution of $\bar{x}$.

Samples from a normal population

It has been shown (p. 111) that the sum of n normal variates is itself a normal variate having a mean equal to the sum of the means and a variance equal to the sum of the variances. Since

$$\bar{x} = (x_1 + x_2 + \cdots + x_n)/n,$$

and the probability distribution of each x is supposedly normal, with mean μ and variance σ^2, it follows that $\bar{x}$ will be distributed normally, with mean $n\mu/n$ ($=\mu$) and variance $n\sigma^2/n^2$ ($=\sigma^2/n$). This agrees with the more general rule mentioned above, and adds the useful fact that $\bar{x}$ is distributed normally.

More surprisingly, it is found that for other types of population, even those which differ widely from the normal, the sample means of large samples tend to be distributed in an approximately normal manner (*see* p. 123, Nos 2, 3)†: so the following methods are of wider application than might at first be supposed.

Example 9.1

A machine is set to cut metal of length 5·40 cm, and its accuracy is such that, whatever the setting, the standard deviation is 0·03 cm. Assuming

† This result is known as the 'Central Limit Theorem'.

that the setting is correct, what is the probability that a sample of 5 rods should give a mean of more than 5·42 cm?

$$\frac{\text{Error of mean}}{\text{S.E. of mean}} = \frac{0\cdot02}{0\cdot03/\sqrt{5}} = 1\cdot49 \text{ approx.}$$

From the table of areas,

$$\text{the probability} = 0\cdot5 - 0\cdot4319,$$
$$= 0\cdot07 \text{ approx, or 1 in 14.}$$

If, in fact, a sample of 5 rods gives a mean of 5·42 cm, the question arises whether this is due to chance or whether it indicates that the setting of the machine is wrong. The above result shows that, if the setting were correct, a mean greater than 5·42 cm is by no means improbable. Indeed, if variations either way are considered there is a chance of 1 in 7 of a deviation greater than the one observed.

Levels of significance

Clearly, it is an arbitrary choice as to what degree of improbability is to be regarded as significant in any particular case. In Example 9.1, the hypothesis being tested (the *null hypothesis*) was that the setting was correct. If the machine had been newly adjusted and the result of a sample were as stated, the probability, 1 in 7, might be sufficient to cast some slight doubt on the accuracy of the setting. At any rate it might be thought that another sample should be taken. But if the machine had been working satisfactorily for some time, notice might not be taken unless the probability was found to be less than 0·05, or possibly less than 0·01. These last two *levels of significance* correspond to values of

$$\left|\frac{\bar{x} - \mu}{\sigma/\sqrt{n}}\right|$$

equal to 1·96 and 2·58 respectively, as can be seen from the table of areas. Results showing greater deviation may be said to be *significant at the 5 per cent level* or *significant at the 1 per cent level*.

Example 9.2

If, in Example 9.1, samples of 5 rods are measured periodically, determine ranges for the mean of the sample such that results outside those ranges would be significant (a) at the 5 per cent level, (b) at the 1 per cent level.

The standard error of the mean $= \dfrac{0\cdot 03}{\sqrt{5}} = 0\cdot 0134$.

For (a), range $= 5\cdot 40 \pm 1\cdot 96 \times 0\cdot 0134$,

i.e. 5·374 to 5·426 cm.

For (b), range $= 5\cdot 40 \pm 2\cdot 58 \times 0\cdot 0134$,

i.e. 5·365 to 5·435 cm.

A control chart, similar to that shown in Fig. 7.1, p. 84, could be designed using these ranges.

The inverse problem

To estimate the mean of a population from the results of a sample is a problem of obvious practical importance but it is necessarily more hazardous than dealing with a population whose characteristics are known. The chief difficulty is that the standard deviation of the population is usually unknown, and the sample may or may not provide a sufficient means of estimating it. When the sample is a large one (say 50 or more) the standard deviation of the sample (s) may be taken as a fair estimate of the population (σ), but for small samples, s and σ may differ widely. The method described below applies only to large samples.

It is necessary to understand the reason for this restriction to large samples. As already explained, $\bar{x}$ is distributed normally if the population from which the sample is drawn is normal, and approximately so in other cases. For samples of size n, since μ and σ are constant, the fraction

$$\frac{\bar{x} - \mu}{\sigma/\sqrt{n}}$$

is also normally distributed. But

$$\frac{\bar{x} - \mu}{s/\sqrt{n}},$$

which we are proposing to use in its place, contains the variable s, and its distribution is far from normal when n is small. The theory of this distribution (called the *t-distribution*) was developed by W. S. Gosset, who wrote under the pseudonym 'Student'. It is considerably more difficult, and we therefore confine our attention to large samples.

Confidence limits

Suppose that a sample of n members ($n \geqslant 50$) gives a mean $\bar{x}$ and standard deviation s. If σ is taken to be equal to s,† and the distribution of $\bar{x}$ is normal, we know that there is a 95 per cent probability that

$$\left|\frac{\mu - \bar{x}}{s/\sqrt{n}}\right| < 1.96,$$

and, hence, that μ will lie in the range

$$\bar{x} - 1.96s/\sqrt{n} \text{ to } \bar{x} + 1.96s/\sqrt{n}.$$

This range is called the *95 per cent confidence interval* for μ, and its boundaries the *95 per cent confidence limits*. Similarly, $\bar{x} \pm 2.58s/\sqrt{n}$ are 99 per cent confidence limits for μ.

Example 9.3

In 1931 there were about 40,000 *railway engine drivers in England and Wales. If a sample of* 120 *gave an average age of* 47·5, *with standard deviation* 9·3, *determine* 95 *per cent and* 98 *per cent confidence limits for the mean age of all the engine drivers.*

$$\text{S.E. of the mean} = \frac{9.3}{\sqrt{(120)}} = 0.85.$$

95 per cent limits are $47.5 \pm 1.96 \times 0.85$, i.e. 45·8 to 49·2. For 98 per cent limits, the table of areas shows that the factor is 2·33. The limits are, therefore, $47.5 \pm 2.33 \times 0.85$, i.e. 45·5 to 49·5.

† It may be remarked that, while s may be either greater or less than σ, according to the particular sample taken, it is rather more likely to be less. This is because s^2 is the mean square deviation measured from the sample mean $\bar{x}$. If measured from μ it would be greater by $(\bar{x} - \mu)^2$. As an estimate of σ^2, s^2 is biased on the low side. Using E for 'expected value',

$$\sigma^2 = E\{s^2 + (\bar{x} - \mu)^2\},$$
$$= E(s^2) + \frac{\sigma^2}{n}.$$

Hence, $$\sigma^2 = \frac{n}{n-1} \times E(s^2).$$

For a given sample, $\frac{n}{n-1} s^2$ is called an *unbiased estimate* of σ^2. It will, however, be seen that for large samples the factor $n/(n-1)$ makes little difference, and may well be ignored.

Difference of means of two large samples

If two samples are taken from the same population, it is not likely that their means will be equal; but neither do we expect them to differ too widely. Suppose that they contain n_1 and n_2 members respectively and that their means are $\bar{x}_1$ and $\bar{x}_2$. If many such pairs of samples were obtained, the difference of their means, $\bar{x}_1 - \bar{x}_2$, would have a frequency distribution centred about zero as a mean value, with a standard deviation which we now proceed to determine. (In terms of a single pair of samples we may say that $\bar{x}_1 - \bar{x}_2$ has a probability distribution with an 'expected value' of zero.) The variances of the distributions of $\bar{x}_1$ and $\bar{x}_2$ are σ^2/n_1 and σ^2/n_2, and it was shown on p. 58 that the difference of two independent variables is distributed about a mean equal to the difference of their means, with a variance equal to the sum of their variances. Hence, the probability distribution of $\bar{x}_1 - \bar{x}_2$ will have an expected value zero, with variance

$$\frac{\sigma^2}{n_1} + \frac{\sigma^2}{n_2}.$$

The standard error of $\bar{x}_1 - \bar{x}_2$ is thus

$$\sigma\sqrt{\left(\frac{1}{n_1} + \frac{1}{n_2}\right)}.$$

We often wish to test whether two samples can reasonably be regarded as coming from the same population. In such a case σ is usually unknown, but can be estimated from the standard deviation of the two samples taken together.† The observed value of $\bar{x}_1 - \bar{x}_2$

---

† If it is desired to estimate σ from the variances s_1^2 and s_2^2 of the two samples σ^2 may be taken as

$$\frac{n_1 s_1^2 + n_2 s_2^2}{n_1 + n_2},$$

provided the samples are large. An 'unbiased' estimate would be

$$\frac{n_1 s_1^2 + n_2 s_2^2}{n_1 + n_2 - 2},$$

for $(n_1 + n_2)\sigma^2 = n_1 . E\{s_1^2 + (\mu - m_1)^2\} + n_2 . E\{s_2^2 + (\mu - m_2)^2$

$$= n_1 . E(s_1^2) + n_1 \frac{\sigma^2}{n_1} + n_2 . E(s_2)^2 + n_2 \frac{\sigma^2}{n_2},$$

$$= n_1 . E(s_1^2) + n_2 . E(s_2^2) + 2\sigma^2.$$

Hence, $$\sigma^2 = \frac{n_1 . E(s_1^2) + n_2 . E(s_2^2)}{n_1 + n_2 + 2}$$

and, for a given pair of samples,

$$\frac{n_1 s_1^2 + n_2 s_2^2}{n_1 + n_2 + 2}$$

is an unbiased estimate of σ^2.

can then be compared with its standard error

$$\sigma\sqrt{\left(\frac{1}{n_1}+\frac{1}{n_2}\right)}.$$

Alternatively (and often more conveniently in practice), the samples may be regarded as coming from two different populations, the question at issue being whether the means of those populations coincide (even though their variances may differ). The standard error of the difference of the means is then

$$\sqrt{\left(\frac{\sigma_1{}^2}{n_1}+\frac{\sigma_2{}^2}{n_2}\right)}.$$

Example 9.4

In an examination, 60 candidates from one school have an average mark of 55, with S.D. 14, and 80 candidates from another school have an average of 52, with S.D. 16. Can it be said that the result obtained by the first school is significantly the better?

The 60 candidates from the first school are regarded as a sample from the largely imaginary population of all such candidates that might have been entered by that school had they been available; and similarly for the second school. The null hypothesis is that these two populations have the same mean, even though their variances may differ. We enquire whether the observed difference, 55 − 52, can be ascribed to chance.

$$\text{S.E. of difference of means} = \sqrt{\left(\frac{14^2}{60}+\frac{16^2}{80}\right)} = 2\cdot54.$$

$$\therefore \frac{\bar{x}_1-\bar{x}_2}{\text{S.E.}} = \frac{3}{2\cdot54} = 1\cdot18 \text{ approx.}$$

The difference is therefore not significant, even at the 5 per cent level. (For a more precise answer, the table of areas shows that the probability of a difference wider than this occurring by chance is 0·238 approx.)

Note. It is important to realize that the conclusion is an entirely negative one. We have failed to prove, by using these figures, that there is a significant difference; but that is not the same as proving that there is no difference. It may indeed be that one of these schools is better, at this sort of examination, than the other. But the figures given fall far short of proving it, or even of making it seem highly probable.

EXERCISE 12

1. Use a table of random numbers to take samples of 5 members from any list of 300 names or more, and record the average number of letters per name for each sample. When 40 or more samples have been taken, make a histogram to show the distribution of the sample means, grouping them 4·5–5·5, 5·5–6·5, etc. Find the mean and standard deviation of this distribution, and fit a normal curve. Estimate the mean and standard deviation of the complete list of names. (Alternatively, use normal probability paper.)

2. The following table shows the averages of 577 bowlers in schools cricket in 1965 (as given in Wisden's *Cricketers' Almanack*):

Average	Frequency	Average	Frequency	Average	Frequency
5–	4	17–	35	29–	3
6–	4	18–	34	30–	4
7–	12	19–	31	31–	3
8–	19	20–	27	32–	3
9–	17	21–	12	33–	0
10–	25	22–	30	34–	3
11–	37	23–	14	35–	3
12–	33	24–	18	36–	2
13–	46	25–	11	37–	0
14–	36	26–	12	38–	0
15–	45	27–	3	39–40	1
16–	46	28–	4		

Make a cumulative frequency table and use a table of random numbers to select samples of 5. Find the means of the samples and make a frequency distribution, grouping them 10·5–11·5, etc.

(Alternatively, use the following means of 40 samples taken in this way:

14·0 15·8 14·8 20·4 13·4 16·2 13·4 15·6 16·4 22·0
23·4 15·6 19·0 14·0 17·4 12·4 15·0 14·2 17·8 15·0
16·2 14·2 19·8 20·4 15·6 11·2 20·8 13·0 17·0 18·4
18·4 14·6 16·2 13·0 13·2 17·0 13·2 15·0 14·2 18·0)

Compare the means and the standard deviations of the original distribution (grouped 5–8, 8–11, etc.) and the distribution formed by the sample means.

3. From a list showing the number of appearances by cricketers in test matches (English players in England, up to 1934) 100 random samples of 5 were taken. The means of the samples were distributed as follows:

Mean	Frequency	Mean	Frequency
1·0–1·8	5	7·0–7·8	7
2·0–2·8	12	8·0–8·8	6
3·0–3·8	21	9·0–9·8	1
4·0–4·8	16	10·0–10·8	0
5·0–5·8	18	11·0–11·8	1
6·0–6·8	13		

The same samples were combined in pairs to form 50 samples of 10, and these gave means as follows:

Mean	Frequency	Mean	Frequency
1·0–1·9	1	5·0–5·9	14
2·0–2·9	8	6·0–6·9	9
3·0–3·9	6	7·0–7·9	2
4·0–4·9	9	8·0–8·9	1

Find the mean and S.D. of each of these distributions, and also of the original distribution, which was as follows:

No. of appearances	Freq.	No. of appearances	Freq.	No. of appearances	Freq.
1	47	8	3	15	0
2	25	9	5	16	2
3	14	10	4	17	1
4	13	11	2	18	1
5	9	12	2	19	1
6	3	13	0	20	2
7	5	14	1	21	1
				22	1

Discuss the results. (*See also* No. 4.)

4. In No. 3 above, the mean of the sample means differs appreciably from that of the population. By treating the mean of the sample means as the mean of 500 separate items, i.e. as the mean of a single 'sample' of 500, show that it does not differ unreasonably from the mean of the population.

5. Castings are mass-produced to have a mean weight of 2·70 kg, and it is found that, over a long period, the standard deviation is 0·03 kg. Find the probabilities
   (i) that a sample of 10 should give a mean weight of more than 2·708 kg,
   (ii) that the mean weight of a sample of 10 should differ from 2·70 kg by more than 0·10 kg.

   If samples of 5 are taken daily, design a control chart to show when the means of samples differ from 2·70 kg by amounts significant (*a*) at the 5 per cent level, (*b*) at the 1 per cent level.

6. A machine is designed to cut lengths of tubing within tolerance limits of 6·22 cm to 6·27 cm. It works to a S.D. 0·01 cm. Assuming normal distribution, find the proportion of its products that will be outside the tolerance limits if it is set to give a mean of (*a*) 6·245 cm, (*b*) 6·250 cm.

   If, when the machine has been working well for some time, with a mean of 6·245 cm, a sample of 20 is taken and the mean of the sample is 6·248 cm, say whether you would advise an adjustment of the setting.

7. A sample of 100 names from a list gave a mean of 6·22 letters, with

S.D. 1·44. Using this standard deviation as an estimate of that of the population, find 95 per cent and 99 per cent confidence limits for the mean of the population.

8. The following table shows the mean number of letters and postcards delivered per morning to boys in two houses at a boarding-school:

House	No. of days	Mean no. delivered	S.D.
C	47	14·5	3·66
D	38	12·6	4·84

Find whether the difference of means is significant.

9. A count was made of the number of letters in the surnames of two batches of examination candidates, with the following results:

	No. of candidates	Mean no. of letters	S.D.
G.C.E.	304	6·49	1·82
Inst. of Actuaries	278	6·95	2·55

Find whether there is a significant difference between the means.

10. Two examiners, marking the same paper, compared results after each had marked 100 scripts. One had a mean of 53 marks, with S.D. 18, and the other a mean of 57, with S.D. 21. Would you say, on this evidence, that one examiner is marking more strictly than the other?

11. Batches of broad-bean seeds were measured in the years 1966 and 1968, and the lengths were distributed as follows:

Length (cm)	Frequencies 1966	1968
1·7	0	1
1·8	1	4
1·9	1	10
2·0	9	29
2·1	9	33
2·2	8	55
2·3	18	39
2·4	16	40
2·5	9	24
2·6	2	6
2·7	0	5
2·8	0	0
2·9	0	1
3·0	0	1
	73	248

Find the mean length for each batch and the mean and standard deviation for the two batches combined. Can the two batches be regarded as having been drawn from the same population?

126   Statistics: the how and the why

12. A group (C) of 50 'careless drivers' were tested over a 12-mile route against a check-group (R) of 50 selected at random. Find on which, if any, of the following indices there is a significant difference between the means for the two groups:

		A/M	V_2	T
Group C	Mean	0·26	42	27·1
	S.D.	0·23	6·3	2·3
Group R	Mean	0·32	39	27·8
	S.D.	0·21	7·9	2·5

(A/M = ratio of number of times mirror was used to number of manœuvres,
V_2 = mean clear speed in mile/hr in derestricted zones,
T = time in minutes to cover the route.)

(Data from S. W. Quenault: *Road Research Laboratory Report* LR70, by permission of the Director of Road Research.)

13. Tests on colour-blindness were made on 66 assorted animals, each animal being given 10 trials at distinguishing between (a) black and white, (b) two other colours. The results were as follows:

No. of successes	Black and white								Colours						
	3	4	5	6	7	8	9	10	2	3	4	5	6	7	8
10 dogs			1	1	2	3	1	2			2	3	4		1
7 cats		2	2			2		1	1			1	3	2	
6 cows				2	3	1					4	5	6		
6 rabbits			1		2		3			1	1	1	3		
8 pigs			1	1	1			2		2	3	2	1		
3 guinea-pigs				1	1	1				1	2				
4 ducks		1		2			1		1			2	1		
Totals	1	5	9	9	8	7		5	3	8	15	14	3		1
22 mice	1	1	4	5	7	2		2	1	2	10	6	3		

(i) Excluding the mice, find the mean number of successes and the S.D. for (a) and for (b).

(ii) Consider the proportion of successes in each of the two series of tests, again excluding the mice. (State the null hypothesis, and compare the observed and expected proportions.)

(iii) Consider the performance of the mice under the same null hypothesis.

(iv) Compare the performance of the mice with that of the other animals. (For a rough comparison the 22 mice may be considered as a 'large' sample.)

(Data from T. J. Gruffydd-Jones.)

14. Measurements of a batch of broad-bean seeds (length in cm) gave the following results:

	Mean
62 from tall parents	2·487
197 from medium parents	2·359
283 from short parents	2·247

The standard deviation for the whole batch was 0·253 cm. Do these figures support the theory that parentage makes no difference?
(Data for Nos. 11 and 14 from Biology VI, Felsted School.)

Errors of observation

A series of observations of any phenomenon is a sample of all possible observations of the same kind. Each observation may be subject to error, and the mean of a series of observed values of a quantity is still only an estimate of the true value. The distribution of the observed values gives some guide to the reliability of the mean and, hence, to that of calculated results obtained by treating the mean as the true value.

First we consider the error in a result due to known errors in the data. This can be found by means of the differential calculus.

If $z = f(x, y, \ldots)$ it follows that

$$\delta z \simeq \frac{\partial f}{\partial x} \delta x + \frac{\partial f}{\partial y} \delta y + \cdots$$

Thus, for small errors, it is a general rule that the error in the result can be found by adding the effects of the separate errors.

If f is a linear function of $x, y, \ldots$, then δz is the same function of $\delta x, \delta y, \ldots$; but for products, quotients and powers it is often convenient to use the fractional error, $\delta z/z$, which can be found approximately by logarithmic differentiation. Thus, if

$$z = x^\alpha y^\beta \ldots,$$
$$\log z = \alpha \log x + \beta \log y + \cdots$$

and

$$\frac{\delta z}{z} \simeq \alpha \frac{\delta x}{x} + \beta \frac{\delta y}{y} + \cdots$$

e.g. if $P = \dfrac{mv^2}{r}$,

$$\frac{\delta P}{P} \simeq \frac{\delta m}{m} + 2 \frac{\delta v}{v} - \frac{\delta r}{r},$$

and if m, v, r are each subject to errors of ± 1 per cent, the maximum error in P is 4 per cent.

But maximum errors are seldom attained, and it becomes necessary to consider the probable combined effect of separate errors whose probability distributions are known. This led to the theory of 'probable errors', i.e. errors whose value is as likely as not to be exceeded. This particular treatment, however, has fallen into disuse, since standard error serves the same purpose more conveniently.

Suppose that $z = f(x_1, x_2, \ldots)$, where $x_1, x_2, \ldots$ are independent variables subject to independent errors, the standard errors of $z, x_1, x_2, \ldots$ being $\sigma, \sigma_1, \sigma_2, \ldots$. It has already been shown (p. 58) that if $z = x_1 \pm x_2 \pm \cdots$, then $\sigma^2 = \sigma_1^2 + \sigma_2^2 + \cdots$.

An easy extension of this result is that if

$$z = k_1 x_1 + k_2 x_2 + \cdots, \quad \text{where } k_1, k_2, \ldots \text{ are constants,}$$

then
$$\delta z = k_1\, \delta x_1 + k_2\, \delta x_2 + \cdots,$$
and
$$\sigma^2 = k_1^2 \sigma_1^2 + k_2^2 \sigma_2^2 + \cdots.$$

In general,

$$\delta z = \left(\frac{\partial f}{\partial x_1}\right) \delta x_1 + \left(\frac{\partial f}{\partial x_1}\right) \delta x_2 + \cdots,$$

and therefore

$$\sigma^2 = \left(\frac{\partial f}{\partial x_1}\right)^2 \sigma_1^2 + \left(\frac{\partial f}{\partial x_2}\right)^2 \sigma_2^2 + \cdots,$$

(since $\partial f/\partial x_1$, etc., are constants, while δx_1, etc., take their various values according to their probability distributions).

Special cases worth noting are:

If $z = x_1 x_2$,

$$\sigma^2 = x_2^2 \sigma_1^2 + x_1^2 \sigma_2^2, \quad \text{and hence} \quad \frac{\sigma}{z} = \sqrt{\left(\frac{\sigma_1^2}{x_1^2} + \frac{\sigma_2^2}{x_2^2}\right)}.$$

If $z = x^n$,

$$\sigma = n x^{n-1} \sigma_1, \quad \text{and hence} \quad \frac{\sigma}{z} = n \frac{\sigma_1}{x}.$$

(σ/z is called the *coefficient of variation* of z.)

In applying these formulae it is, of course, necessary to know the values of σ_1, σ_2, etc., and this is usually a matter of sampling. If n independent measurements are made of a quantity x, it is best, as explained in the footnote on p. 120, to use the 'unbiased estimate'

$$\frac{n}{n-1} s^2,$$

where s^2 is the variance of the n measurements. If n is small, however, as often happens in scientific experiments, it must be realized that

Sampling

δz may not be distributed normally, and the table of areas under the normal curve cannot properly be used for the interpretation of the resulting value of σ. Small samples can, however, be dealt with by means of the t-test, as described in books on statistical method. (The theory of this test is more difficult.)

In scientific work it is always desirable to indicate the degree of accuracy of any measurement, but the practice of writing a value as, for example, $14\cdot 2 \pm 0\cdot 05$ cannot be recommended, since different writers have used this notation variously to indicate the maximum error, the standard error, and the so-called 'probable error'. Unless there is a clear understanding as to the meaning it is better to say '14·2, with S.E. 0·05' or '14·2, with max. error 0·05'.

Example 9.5

Telephone poles are placed along a road at intervals whose mean is $\frac{1}{28}$ mile, with standard deviation $\frac{1}{2}$ yard. A motorist wishing to check his speedometer times himself over a supposed $\frac{1}{4}$ mile, as measured by 7 of these intervals. His reading is 30 seconds and is subject to a standard error of $\frac{1}{10}$ sec. Estimate his speed in mile/hr, stating the standard error.

If dist. $= x$ yd, time $= t$ sec, and speed $= v$ mile/hr,

$$v = \frac{x}{1760} \times \frac{3600}{t} = \frac{90x}{44t}.$$

$$\therefore \sigma_v^2 = \left(\frac{90}{44t}\right)^2 \sigma_x^2 + \left(\frac{90x}{44t^2}\right)^2 \sigma_t^2,$$

$$= \frac{v^2}{x^2}\sigma_x^2 + \frac{v^2}{t^2}\sigma_t^2,$$

where $t = 30$, $\sigma_t = \frac{1}{10}$, $x = 440$ and $\sigma_x^2 = 7 \times (\frac{1}{2})^2$.

Hence, $v = 30$ and $\sigma_v^2 = 0\cdot 01815$. The speed is therefore 30 mile/hr, with a S.E. of 0·135 mile/hr.

(The working may be shortened by noticing that

$$\log v = \text{const.} + \log x - \log t,$$

and, hence,

$$\frac{1}{v}\frac{\partial v}{\partial x} = \frac{1}{x}, \quad \frac{1}{v}\frac{\partial v}{\partial t} = \frac{1}{t}.\Big)$$

Errors due to approximation

When a quantity is known to a given degree of accuracy (e.g. to 1 decimal place) it is to be supposed that the probability distribution of the error is uniform over the range (± 0.05 in such a case). It was proved on p. 66 that, if the range is $\pm \epsilon$, the variance is $\epsilon^2/3$. Hence, if n such quantities are added together, the variance of the result is $n\epsilon^2/3$, and the standard error $\epsilon\sqrt{(n/3)}$. If the average is taken, its S.E. is $\epsilon/\sqrt{(3n)}$. Similarly if α, β, γ are all subject to errors of $\pm \epsilon$, uniformly distributed and independent of one another, the variance of $l\alpha + m\beta - n\gamma$ is $(l^2 + m^2 + n^2)\epsilon^2/3$; and so on.

Example 9.6

Find the coefficient of variation of the geometric mean of 10 quantities, each of which is known within $\tfrac{1}{2}$ per cent.

If m is the required mean,

$$m^{10} = \alpha\beta\ldots$$
$$10 \log m = \log \alpha + \log \beta + \cdots$$
$$\frac{10}{m} \delta m = \frac{1}{\alpha} \delta\alpha + \frac{1}{\beta} \delta\beta + \cdots$$
$$\therefore \left(\frac{10}{m}\right)^2 \sigma^2 = \left(\frac{1}{\alpha}\right)^2 \frac{\epsilon^2}{3} + \left(\frac{1}{\beta}\right)^2 \frac{\epsilon^2}{3} + \cdots$$

Therefore the proportional S.E. (or coefficient of variation),

$$\frac{\sigma}{m} = \frac{1}{10} \sqrt{\left(\frac{10}{3 \times 200^2}\right)} = \frac{1}{200\sqrt{30}}.$$

EXERCISE 13

1. If $E = \tfrac{1}{2}mv^2$, find the approximate error in E when $m = 10$, with error 0.15, and $v = 30$, with error 0.5.
   Find also the maximum percentage error in E if m and v are each subject to errors of ± 0.5 per cent.
2. If $V = \sqrt{(u^2 + v^2)}$ and $u = 200$, $v = 150$, u and v each being subject to a maximum error of ± 0.5, find the maximum error in the calculated value of V.
   If the standard errors of u and v are each $0.5/\sqrt{3}$, find the S.E. of V.
3. If $R = PV/T$, find the maximum percentage error in R due to errors of $\pm\tfrac{1}{2}$ per cent in each of the other quantities.
   If the coefficients of variation of P, V, and T are 0.002, 0.005, and 0.003 respectively, find that of R.
4. A runner is timed over 400 metres on many occasions, and his mean time is 53.6 seconds, with standard deviation 1.4 sec. Find his mean average speed in m/sec, and its standard deviation.

5. The range of a projectile on a horizontal plane is given by $R = (V^2/g) \sin 2\alpha$, where V is the initial velocity and α the angle of elevation. If $V = 100$, with S.E. 1·25, $\alpha = 30°$, with S.E. $\frac{1}{2}°$, and $g = 32·2$, find R and its S.E.

6. Find the standard error of (a) the sum, (b) the average, of 60 numbers, each correct to the nearest unit.

7. An experimenter makes ten independent measurements of a physical quantity, with the following results:

$$+0·13, \quad +0·05, \quad +0·08, \quad +0·02, \quad +0·10,$$
$$-0·01, \quad +0·04, \quad +0·05, \quad +0·09, \quad +0·05$$

Make an 'unbiased estimate' of the variance of such results and use it to estimate the S.E. if he were to take the mean of 50 such observations.

Sampling of attributes

We are often concerned with events whose outcome is one of two possibilities, which may be conveniently labelled 'success' or 'failure'. In a single trial the number of successes is either 0 or 1, and we suppose that the corresponding probabilities are q and p (both constant). These are the probabilities in taking a sample of one from a population in which the proportion of failures to successes is $q:p$. The probability distribution is as shown in Fig. 9.1. If n trials are made, they constitute a random sample of n members from the same population. It has been shown (pp. 45, 75) that the distribution of the number of successes in such samples is binomial, with a mean np and a standard deviation $\sqrt{(npq)}$. In terms of the proportion of successes the mean or 'expected' proportion is p, and the standard deviation $\sqrt{(pq/n)}$.

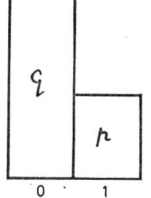

Fig. 9.1

It has also been shown (p. 109) that, when n is large, the binomial distribution approximates to the normal, and we are then able to estimate the probability of the sample's containing more than a given number of successes by using the table of areas under the normal curve. Thus, if a die, supposed true, is thrown 120 times, the expected number of 'sixes' is 20, with S.D. $\sqrt{(120 \times \frac{1}{6} \times \frac{5}{6})}$, or 4·1 approx. If the observed number of 'sixes' is 32, we say it is 'significantly large', because it exceeds the expected number by 12, which is about 2·93 times the S.D. From the table of areas the probability of an excess as great as this occurring by chance is $0·5 - 0·4983$, or 0·0017 (about 1 in 590). On this result we should doubt the hypothesis that the die is true.

How large n must be for the estimate by means of the normal curve to be sufficiently good for practical purposes is a matter of judgement, but some guide is given by the diagrams of Figs 8.1, 8.2. and 8.3 (pp. 92, 94 and 97). As in Example 8.2 (p. 97) it is better, when n is fairly small (say 10 to 40), to replace x/σ by $(x - \frac{1}{2})/\sigma$ or $(x + \frac{1}{2})/\sigma$, according to which side the area is being measured.

Levels of significance

As already mentioned, the degree of improbability to be regarded as significant is a matter of arbitrary choice. A probability of under 5 per cent is enough to cause doubt, but is far from conclusive. A probability of under 1 per cent is much stronger evidence against the hypothesis, and one of less than 0·1 per cent is almost conclusive. The meaning of the term 'significance' is made more precise by the use of such phrases as 'significant at the 5 per cent level'.

Example 9.7

Suppose it is said that one person in 5 reads the 'Daily Disaster'. If among 50 people chosen at random there are only 6 who read it, is that evidence against the assertion?

The deficiency below the expected number is 4, and the standard deviation is $\sqrt{(50 \times \frac{1}{5} \times \frac{4}{5})}$, or 2·83. Thus, $X/\sigma = 4/(2·83) = 1·41$. From the table of areas, the chance of a deficiency greater than this is $0·5 - 0·4207$, or 0·08 approx. This result is not significant, even at the 5 per cent level, so the assertion cannot be contradicted on this evidence alone.

(If, for a more accurate estimate, X is taken as 3·5 instead of 4, the result is 0·11 approx., and the conclusion is the same. The discrepancy between the two methods may seem considerable but, as is evident from the shape of the normal curve, it would be much less for the higher, more crucial, values of X/σ.)

Note. It is sometimes difficult to know whether to consider the probability of deviations in one or both directions from the expected value. In this example it seems natural to consider only the possibility of deficiencies. The question is whether the estimate of 1 in 5 is too high. Thus we consider only one 'tail' of the curve. If the question were, 'Is the estimate an accurate one?' we should consider two 'tails', representing deviation either way, and the probability would be twice as great.

The inverse problem

As before, we deal only with large samples.

Suppose it is found that a sample of 100 members yields 30 successes and 70 failures. What can be inferred about the proportion p of successes in the population? If a large number of samples of 100 were taken, their means $\bar{x}$ would be distributed about the expected values $100p$ with S.D. $\sqrt{(100pq)}$, and the distribution would be approximately normal. In about 95 per cent of such samples the mean of the sample would lie in the range $100p \pm 1.96\sqrt{(100pq)}$. For a first approximation to $1.96\sqrt{(100pq)}$ we may take p as 0.3 and q as 0.7, giving $1.96\sqrt{(21)}$, or 9 approx. For the observed value 30 to lie within the range $100p \pm 9$, p must be between 0.21 and 0.39. These are called *95 per cent confidence limits* for p.

(For a closer approximation it may be noted that, if p were as small as 0.21, $1.96\sqrt{(100pq)}$ would be 8 approx., and if p were as large as 0.39 it would be nearly 10. The confidence limits would then be 0.22 and 0.40.)

If 99 per cent confidence limits are desired, we find from the table of areas that, when the two 'tails' of the curve have an area of 0.005 each, $X/\sigma = \pm 2.58$. Thus, in 99 cases out of 100, the mean of the sample, 30, should lie within the range $100p \pm 2.58\sqrt{(100pq)}$, i.e. $100p \pm 11.8$ approx. From this, the 99 per cent confidence limits for p are approximately 0.18 and 0.42. (The closer approximation gives 0.20 and 0.426.)

Comparison of two large samples

If two samples, of n_1 and n_2 members respectively, are taken from a population in which the probability of 'success' is p, the expected proportion of successes is the same for both, namely p. The variances are pq/n_1 and pq/n_2. The difference between the proportion of successes in the two samples will therefore have an expected value zero, with variance

$$\frac{pq}{n_1} + \frac{pq}{n_2}.$$

The S.E. of the difference is thus

$$\sqrt{\left\{pq\left(\frac{1}{n_1} + \frac{1}{n_2}\right)\right\}}$$

In practice, p must usually be estimated as

$$\frac{\text{total number of successes in the two samples}}{n_1 + n_2}.$$

Example 9.8

Of two examiners for driving tests, one passed 84 *candidates out of* 150, *and the other* 120 *out of* 200. *Do these figures indicate that one was stricter than the other?*

Taking p as the proportion passed, i.e. 204/350, or 0·583, we consider whether the difference in the proportions passed by the two examiners could be due to the chance of sampling.

$$\text{S.E. of difference} = 0{\cdot}583 \times 0{\cdot}417 \sqrt{\left(\frac{1}{150} + \frac{1}{200}\right)},$$
$$= 0{\cdot}0262.$$

Observed difference = 0·60 − 0·56 = 0·04.

$$\text{Ratio } \frac{\text{observed difference}}{\text{S.E.}} = \frac{0{\cdot}04}{0{\cdot}0262} = 1{\cdot}53 \text{ approx.}$$

Applying the 'two-tailed' test, the table of areas under the normal curve shows that the probability of a greater difference (either way) occurring by chance is 0·126, or about 1 in 8. The figures thus provide no evidence of a difference in standard between the two examiners. (Nor do they prove that there is no difference. The conclusion is an entirely negative one.)

Simple sampling

It has been assumed above that both n and p are constant. When we say that the variance for a sample of size n is npq, we mean that if a large number of samples *of the same size* were taken, they would show such a variance. We are further supposing that the probability p of success is the same for all samples and for all members of the same sample. This is called *simple sampling*.

If six cards are drawn from a pack without replacement, the probability of drawing an ace varies with each card drawn. This is not simple sampling. If a large number of men fire at a target and it is hit, on the average, 9 times out of 10, it is not to be supposed that each man, when he fires a shot, has a 9/10 chance of success, because their individual skills may vary; so if the samples consist of 20 shots each, by various particular men, p varies from sample to sample and it is not simple sampling. Again, if samples of sometimes 9, sometimes 10, and sometimes 11 items are taken from the products of a manufacturing process, the number of rejects will vary more than if all the samples were of 10 items. This, again, is not simple sampling.

If the samples are drawn at random from an infinite (or very large) population, it is fair to suppose that p is constant; but if the population is finite, p is not constant unless the members of the sample are drawn one by one, each being 'replaced' before the next is drawn.

*Sampling without replacement from a finite population

If a sample of n members, all different, is taken from a population of N members, the variance of the sample mean is not σ^2/n but

$$\frac{N-n}{N-1}\cdot\frac{\sigma^2}{n}.$$

(If N is large, σ^2/n is a good approximation.)

Proof: It is convenient to measure from the mean of the population as origin. Let $x_1, x_2, \ldots, x_N$ be the values so measured (so that $\sum x = 0$). If z is the sum of n of these, e.g. $x_1, x_2, \ldots, x_n$, z will take $_NC_n$ possible values, each x appearing $_{N-1}C_{n-1}$ times. Thus,

$$\bar{z} = \frac{_{N-1}C_{n-1}}{_NC_n}\sum x = 0.$$

Moreover, in the $_NC_n$ values of $(x_1 + x_2 + \cdots + x_n)^2$, each x^2 will appear $_{N-1}C_{n-1}$ times and each product $2x_ix_j$ will appear $_{N-2}C_{n-2}$ times. But

$$x_1(x_2 + x_3 + \cdots + x_N) = x_1(N\bar{x} - x_1) = -x_1^2;$$
$$x_2(x_1 + x_3 + \cdots + x_N) = x_2(N\bar{x} - x_2) = -x_2^2;$$

and so on.

Hence, $$\sum 2x_ix_j = -\sum x_i^2,$$

and

the sum of all such product terms $= {}_{N-2}C_{n-2}\sum 2x_ix_j,$
$$= -{}_{N-2}C_{n-2}\sum x_i^2.$$

The variance of $z = \dfrac{\sum(x_1 + x_2 + \cdots + x_n)^2}{_NC_n},$

$$= \frac{_{N-1}C_{n-1}\sum x^2 - {}_{N-2}C_{n-2}\sum x^2}{_NC_n},$$

$$= \frac{_{N-1}C_{n-1} - {}_{N-2}C_{n-2}}{_NC_n}\cdot N\sigma^2,$$

$$= \frac{\dfrac{N-1}{n-1} - 1}{\dfrac{N(N-1)}{n(n-1)}}\cdot N\sigma^2,$$

$$= \frac{N-n}{N-1}\cdot n\sigma^2.$$

For example: the numbers 0, 1, 2, 3, ..., 99 have a mean value of 49·5, with variance 833·25. If ten of these numbers are selected by means of a table of random numbers, the sample so obtained may sometimes contain the same number twice, or even three times. This is simple sampling, and the standard error of the mean will be $\sqrt{(833\cdot 25/10)}$, or 9·13. But if ten of the numbers, *all different*, are selected (e.g. by picking from a table of random numbers with the proviso that repetitions are to be rejected), the standard error of the sample mean will be

$$\sqrt{\left(\frac{90}{99} \times \frac{833\cdot 25}{10}\right)}, \quad \text{or} \quad 8\cdot 70.$$

*Samples of varying size

Suppose that the size of the sample is a variable number n whose mean is $\bar{n}$ and whose variance is σ_n^2, and that the probability of success for any member of a sample is p (constant). For samples of a particular size n the mean number of successes is np and the mean square deviation from np is npq (where $q = 1 - p$). For all the samples together, the mean of the sample means is $\bar{n}p$. For the variance we therefore have to find the mean square deviation from $\bar{n}p$. For samples of size n, this is $npq + (np - \bar{n}p)^2$, or $npq + (n - \bar{n})^2 p^2$, and the variance is the expected value of this, namely $\bar{n}pq + \sigma_n^2 p^2$.

For example, suppose that a bowl contains a large number of beads, of which a quarter is red, well mixed together, and that cupfuls are taken, the mean number in a cupful being 100, with standard deviation 5. The mean number of red beads in a cupful will be 25, with standard deviation

$$\sqrt{(100 \times \tfrac{1}{4} \times \tfrac{3}{4} + 25 \times \tfrac{1}{16})}, \quad \text{or} \quad 4\cdot 51.$$

EXERCISE 14

1. Using a table of random numbers, six boys each picked 50 pairs of two-figure numbers and reported that 203 out of 300 pairs were prime pairs. The theoretical proportion of prime pairs is approximately $6/\pi^2$ ($=0\cdot 608$). Is it probable that their method was faulty?
2. A test on a sample of 100 seeds showed a germination rate of 75 per cent. 200 seeds from a second consignment showed a rate of only 70 per cent. Can the two samples be regarded as coming from the same population? State the probability that the difference could be due to chance.
3. Of two driving-test examiners, testing candidates drawn at random

from the same population, one passes 44 out of 80, and the other 57 out of 120. Is there reason to suppose that one is stricter than the other?

4. Of 1226 railway accidents in 1956 it was found that 640 were due to human failure; the corresponding figures for 1957 were 1205 and 621. (i) Is there a significant improvement in the proportion due to human failure? State the probability of such an improvement occurring by chance. (ii) If on a particular line there were 30 accidents, of which 20 were due to human failure, find the probability that such a high proportion might be due to chance.

5. In an opinion poll, in which 1000 people were questioned, 38 per cent stated that they had no confidence in the Government's economic policy. Find confidence limits (*a*) at the 95 per cent level, (*b*) at the 99 per cent level, for the percentage of the population taking that view.

6. In a manufacturing process, a series of 20 samples, of 10 items each, gave an overall proportion of 8 per cent of items which did not pass a certain stringent test. Find 95 per cent confidence limits for the percentage of defective items.

If, after adjustments have been made to the machinery, a further series of 12 samples, of 10 items each, gave a proportion of 5 per cent defectives, find the probability that the improvement could be due to chance. What further action, if any, would you suggest?

*7. A piano firm has a stock of 100 instruments whose mean value is £150, with S.D. £20. If 40 are selected at random to stock a new branch establishment, find 95 per cent confidence limits for their mean value.

*8. The 'point-count' of a bridge hand is made by counting 4 for an ace, 3 for a king, 2 for a queen and 1 for a jack. Show that, if a single card is drawn from a pack, its mean point-count is 10/13, with variance 290/169; and that the mean point-count for a hand of 13 cards is 10, with variance 290/17.

Can this be used for estimating the probability of a hand's having a point-count of over 20?

Notation and formulae

Mean and variance

If $X_1, X_2, \ldots, X_n$ are n values of a variable X,

$$\text{the mean } \bar{X} = \frac{\Sigma X}{n};$$

the variance (or second moment about the mean)

$$s_X^2 = \frac{\Sigma (X - \bar{X})^2}{n} = \frac{\Sigma X^2}{n} - \bar{X}^2;$$

and the standard deviation $= \sqrt{\text{(variance)}} = s_X$.

Change of origin and scale

If $X = a + kx$, where a is an 'assumed mean' and x is the 'working variable',

$$\text{the mean } \bar{X} = a + k\bar{x};$$

and the variance $s_X^2 = k^2 s_x^2 = k^2 \left(\frac{\Sigma x^2}{n} - \bar{x}^2 \right).$

Frequency distributions

If $X_1, X_2, \ldots,$ occur with frequencies $f_1, f_2, \ldots,$

$$\bar{X} = \frac{\Sigma fX}{n} \quad \text{and} \quad \bar{x} = \frac{\Sigma fx}{n};$$

$$s_X^2 = \frac{\Sigma fX^2}{n} - \bar{X}^2 \quad \text{and} \quad s_x^2 = \frac{\Sigma fX^2}{n} - \bar{x}^2.$$

$\bar{X} = a + k\bar{x}$ and $s_X = k s_x$, as before.

Continuous distributions

If $f(x)$ is the frequency-density for a variable x,

$$\text{the total frequency } n = \int f(x)\, dx;$$

$$\text{the mean } \bar{x} = \frac{1}{n} \int x . f(x)\, dx;$$

and

$$\text{the variance } s^2 = \frac{1}{n} \int (x - \bar{x})^2 . f(x)\, dx = \frac{1}{n} \int x^2 . f(x)\, dx - \bar{x}^2.$$

Bivariate distributions

If $(X_1, Y_1), \ldots, (X_n, Y_n)$ are n pairs of values of variables X, Y, and $(x_1, y_1), \ldots, (x_n, y_n)$ are those of the 'working variables' x, y, where $X = a + kx$ and $Y = b + ly$, the covariance, or product-moment about $(\bar{X}, \bar{Y})$,

$$p_{X,Y} = \frac{\sum (X - \bar{X})(Y - \bar{Y})}{n},$$

$$= kl \frac{\sum (x - \bar{x})(y - \bar{y})}{n} = kl \left(\frac{\sum xy}{n} - \bar{x}\bar{y} \right).$$

The regression line of Y on X is

$$Y - \bar{Y} = \frac{p_{X,Y}}{\sigma_X^2} (X - \bar{X}).$$

$$\text{Correlation coefficient} = \frac{p_{XY}}{\sigma_X \sigma_Y} = \frac{p_{xy}}{\sigma_x \sigma_y}.$$

Probability distributions

If $p_1, p_2, \ldots$, are the probabilities of a variable taking values $x_1, x_2, \ldots$, where $p_1 + p_2 + \cdots = 1$,

$$\text{the expected value } E(x) = \mu = \sum px;$$

and

$$\text{the variance } \sigma^2 = \frac{\sum px^2}{n} - \mu^2.$$

If $f(x)$ is the probability-density for a continuous variable x,

$$\text{the total probability} = \int f(x)\, dx = 1;$$

the expected value $E(x) = \mu = \int x f(x)\,dx;$

and the variance $\sigma^2 = \int x^2 f(x)\,dx - \mu^2,$

the integrals being taken over the possible range of values of x.

Binomial probability distribution
If p is the probability of success and q that of failure in any single trial, $p + q = 1$, and

the probability of r successes in n trials $P(r) = {}_nC_r q^{n-r} p^r$;
expected number of successes $= np$;
variance $= npq$; and standard deviation $= \sqrt{(npq)}$.

The expected proportion of successes $= p$, with variance pq/n, and standard deviation $\sqrt{(pq/n)}$.

Poisson probability distribution
If m is the expected number of successes ($= np$, where n is large and p small),

$$P(r) = e^{-m} \cdot \frac{m^r}{r!};\quad \text{variance} = m;\quad \text{and standard deviation} = \sqrt{m}.$$

Normal probability distribution
If m is the expected value of a variable X and σ the standard deviation,

the probability-density $y = \dfrac{1}{\sqrt{(2\pi\sigma^2)}}\, e^{-(X-m)^2/(2\sigma^2)};$

or, if $X = m + x$,

$$y = \frac{1}{\sqrt{(2\pi\sigma^2)}}\, e^{-x^2/(2\sigma^2)}.$$

Sampling
If a population consists of N members, with mean μ and variance σ^2, and a sample of n members, $x_1, x_2, \ldots, x_n$, is taken, with mean $\bar{x}$ and variance s^2,

the mean of the sample has a probability-distribution having a mean m and variance σ^2/n.
Standard error of the mean $= \sigma/\sqrt{n}$.

From a large sample, estimate of mean of the population $= \bar{x}$, and unbiased estimate of variance $= ns^2/(n-1)$.

Answers

Exercise 1 (p. 19)
1. (i) Bar diagram. (ii) Pie diagram. (iii) Double bar diagram. (iv) Time chart. (v) Histogram.
2. No; effect of Christmas trade and of February being a short month.
3. 28·1, 20·4 per acre; boundaries extended.
4. 16·7, 16·5; both did badly, the first in a bigger way.
5. Less; the longer books have a greater chance of appearing on the day.
6. Assuming that King's figures were correct, the proportion was probably higher than one-tenth, since the 600,000 were probably survivors of a smaller population, which had, moreover, been considerably reduced during the plague years.
7. Mode 12 clicks; median 11; L.Q. 9, U.Q. 13; S.I.Q.R. 2.
8. Median 146, L.Q. 128, U.Q. 159, S.I.Q.R. 15·5 (all in lb).
9. (i) Median 44, L.Q. 35·5, U.Q. 52·5, S.I.Q.R. 8·5; (ii) 45; (iii) 38.
10. 1·55, 0·35.
11. 2·70 cm, 0·216 cm.
12. 6.2, 1·48 microns.
13. 45, 11·4; 4·8 per cent outside.
14. 26·4, 14·9(5).
15. 36·3, 9·95.
16. 69·1 in, 2·60 in.
17. Plot points on graph paper to illustrate the figures and draw through them a smooth S-shaped curve. From this curve (which is a cumulative frequency curve) make a cumulative frequency table at intervals of 0·01 or 0·025 concentrations. (It is sufficient to work to the nearest 1 per cent.) Hence, find the frequencies for the corresponding intervals, and calculate the approximate mean concentration. 0·674.
18. S.I.Q.R. 22, giving S.D. 33; S.D. = 29·0.
19. 3·02, 1·33, 5 per cent.
21. 7·18.
22. 41 mm.
23. 15·7.
24. Mean $= \dfrac{n_1 m_1 + n_2 m_2}{n_1 + n_2}$.
25. (i) 13·1, 1·69, (ii) 14·3, 2·06, (iii) 27·4, 3·75.

Exercise 2 (p. 34)
1. (i) 50, 1·22, (ii) 400, (iii) 8·7, (iv) $y = 0·022x + 0·13$.
2. $y = 0·407 - 21·8$.
3. $y = 0·3x - 1·23$.
4. $Y = 2·47X - 286$; 7·4, 12·3.

Answers 143

5. (i) 11·37, (ii) $Y = 3\cdot25X - 7\cdot5$, (iii) 0·94.
6. 0·97.
7. $Y = 0\cdot744X + 26\cdot3$.
8. 0·69.
9. $-0\cdot28$.

Exercise 3 (p. 46)

1. (i) 3^{10}, (ii) 720, (iii) 120.
2. 3060.
3. 1452.
4. 180, (i) 60, (ii) 15.
5. $\dfrac{n!}{p!q!r!}$; 835.
6. (i) $_{22}C_{11} = 705\,432$, (ii) 352 716.
7. 255.
8. 255, 282 240.
9. (i) $9! = 362\,880$, (ii) 60 480.
10. 132, 462, 495, 27 720, 720, 10 395, 4094.
12. (i) 15, 8, (ii) 105, 60, (iii) 1365, 330, 35, 35.
13. (i) 120, (ii) 20, (iii) 60, (iv) 10, (v) 10.
16. $\frac{1}{36}, \frac{5}{36}$.
17. (i) 0·125, (ii) 0·225, (iii) 0·18; 0·84.
18. $\frac{5}{28}, \frac{125}{512}$,
19. $\frac{5}{24}, \frac{1}{2}$.
20. (i) $\frac{3}{20}$, (ii) $\frac{1}{6}$.
21. $p^2 + 2pp' - 3p^2p'$.
22. (i) 4850, (ii) $\frac{1}{5}$, (iii) $\frac{3}{190}$, (iv) $\frac{32}{625}$.
23. $\frac{8}{21}$.
24. (i) $\frac{29}{59}$, (ii) $\frac{1}{3}$.
25. $\frac{1}{5}$.
26. $\dfrac{1}{27}, \dfrac{26}{27^5}$.
27. $\frac{61}{216}, \frac{625}{1296}$.
28. (i) $\frac{5}{17}$, (ii) $\frac{5}{68}$, (iii) $\frac{51}{136}$.
29. (i) $\dfrac{3}{5}$, (ii) $\dfrac{10n - 1}{18n - 3}$.
30. (i) $\frac{3}{5}$, (ii) $\frac{11}{15}$.
31. 0·04552. (i) 0·00048, (ii) 0·04504.
32. $(\frac{1}{4} + \frac{3}{4})^5$. (i) $\frac{45}{512}$, (ii) $\frac{53}{512}$.
33. $\frac{280}{2187}, \frac{11}{243}$.
34. 17:3.
35. $\frac{32}{663}$; odds 428:1 against.
36. Sum of chances $= \frac{251}{180}$.
37. 0·073, 0·25.
38. 0·57.
39. 0·214, 0·195; no.
40. (i) $\frac{1}{4096}$, (ii) $\frac{1}{512}$, (iii) $\frac{79}{4096}$, (iv) $\frac{299}{4096}$ (or 1 in 13·7).

Exercise 4 (p. 54)

1. £10·25.
2. 2·25s.

3. (i) $1\frac{1}{6}s$., (ii) $2·46s$. approx.
4. $2\frac{2}{3}s$., $8s$.
5. £480·1.
6. $\frac{1}{32}, \frac{5}{32}, \frac{10}{32}, \frac{10}{32}, \frac{5}{32}, \frac{1}{32}; 2\frac{1}{2}$.
7. $\frac{1}{3^6}, \frac{12}{3^6}, \frac{60}{3^6}, \frac{160}{3^6}, \frac{240}{3^6}, \frac{192}{3^6}, \frac{64}{3^6}; 4$.
8. 118, 10·6.

Exercise 5 (p. 63)

1. 11 ft 1 in, 3·40 in; 5 ft 6½ in, 1·70 in.
2. Statics 22·9, 10·1; dynamics 19·6, 8·81; 42·5, 13·4 min.
3. 31·25, 5·59 min; 173·2, 206·8 min.
4. 2·79 mile/hr.
5. No (only 1·3 × S.E.).
6. Yes (over 3 × S.E.).
7. (i) Yes (nearly 3 times), (ii) No (less than twice).
8. (i) Another test (nearly twice S.E.). (ii) Investigate (nearly 3 times S.E.).
9. −2·81 units, 2·5 units; no action.
10. No (less than twice S.E.).
11. 45·18 to 45·48 g.
12. 1·13 min; No.
13. 13.

Exercise 6 (p. 69)

1. 80, 0·45.
2. 6, 2.
3. 4, 2·16.
4. $1/k$, $1/k$, $1 - 1/e$.
5. Var. = $\frac{2}{3}$.
6. Possible error of 1 in 4th place.
7. (i) Error not often more than £3, (ii) Not often more than £0·1.
9. 9·5 min, 3·23 min; 3·0 to 16·0 min.
10. 3 miles, 0·41 mile.
11. Groups 33–37, etc., 209·9, 207·8
    34–38   189·8, 187·7
    35–39   190·3, 188·2
    36–40   189·4, 187·3
    37–41   200·1, 198·0
(Average of corrected values 193·8; true value 193·9.)

Exercise 7 (p. 78)

1. (i) $\frac{5}{16}$, (ii) $\frac{11}{2048}$, (iii) $\frac{4089}{4096}$.
2. 0·358, 0·55.
3. 1 in 50.
4. $22\frac{1}{9}$; No.
5. No (S.D. 31·8).
6. Yes (more than 3 times S.D.); Yes (together, excess $\simeq$ 5 times S.D.).
7. Excess 0·002; not large (less than twice S.D.).
8. Yes (over 4 times S.D.).

Answers 145

9. 0·72.
10. Excess 28·9 (=1·74 × S.D.); No.
11. 2·4 per cent.
12. 72·5; not significant (excess ≃ 1·26 × S.D.).
13. Yes (deficiency more than 6 times S.D.).
14. No (1·25 times S.D.); Yes (2·55 times).
15. 300.
16. 2, 0·294; 20·3.
17. 0·79.
18. Yes (excess less than S.D.); No (excess ≃ 1·5 × S.D.).
19. 3·7 × 10^6.

Exercise 8 (p. 85)

1. 90·4, 108·4, 65·1, 26·0, 7·8, 1·9, 0·4, 0·1; 0·034.
2. A 0·844; 38·7, 32·6, 13·8, 3·9, 0·8, 0·1.
   B 2·06; 11·5, 23·7, 24·4, 16·7, 8·6, 3·5, 1·2, 0·3.
3. (i) 1, (ii) 12·9.
4. 0·018.
5. (a) 0·037, (b) 0·0037; (i) Another sample, (ii) Look for faults.
6. (i) 0·082; no action. (ii) 0·0067; see if tests are being properly done.
7. (i) 0·533, (ii) 0·132.
8. (i) Yes (nearly 5 times S.D.) (not Poissonian). (ii) Yes (probability only 1 in 164).
9. Yes (probability of happening by chance about 1 in 910).

Exercise 9 (p. 88)

1. (i) 0·065, (ii) 0·27.
2. (i) 0·10, (ii) 0·001, (iii) 0·008.
3. (i) 0·07, (ii) 0·012.
4. (i) 0·16, (ii) 0·24; 0·001 (binomial, not Poisson).
5. No (probability of 5 or more 0·23); low (probability of 6 or less 0·001).
6. Epidemic likely (probability of 3 = 0·007); almost certain (0·0007).
7. 0·929; 227·5; 211·3, 98·2, 30·4, 7·1, 1·3.
8. 14.
9. 10.
10. Frequencies 4, 15, 17, 25, 17, 8, 1, 2, 0, 1; expected values 4·9, 14·3, 20·8, 20·1, 14·6, 8·5, 4·1, 1·7.

Exercise 10 (p. 100)

1. 0·9 per cent; 1854 hr or more.
2. 83 per cent, 4·4 per cent.
3. 348; above 113·5.
4. 21·7, 13·9 per cent.
5. 11·5 per cent; 0·954.
6. 0·067, 0·114.
7. (i) 0·048, (ii) 0·006, (iii) Yes (deviation nearly 3 × S.E.).
8. (i) 0·048, (ii) 0·067, (iv) 0·25.
9. 0·079.
10. (i) 0·006, (ii) 0·048, (iii) 0·083, (iv) 0·42.

146   Statistics: the how and the why

11. (i) 844, (ii) 63·5, (iii) No (probability of result by chance = 0·055).
12. (i) 5·0, 2·4 per cent, (ii) 7·3 per cent, (iii) 0·27.
13. 0·024 cm; 22 per cent.
14. 50·3, 19·2 per cent; 30·7 per cent.
15. 0·39.

Exercise 11 (p. 107)

1. Exp. $f.$: 4·6, 9·8, 17·4, 28·6, 38·9, 44·8, 45·5, 39·7, 29·8, 19·2, 10·7, 5·1, 2·1, 0·8, 0·3.
2. 9·4, 31·3, 57·6, 57·6, 31·3, 9·4.
3. 5·8, 12·7, 23·0, 34·2, 42·1, 41·9, 36·2, 25·3, 14·6, 6·9, 2·7, 0·9.
4. 109, 11·1.
6. 1·0, 3·0, 7·4, 14·9, 25·4, 33·6, 38·1, 33·6, 25·4, 14·9, 7·4, 3·0, 1·0.
7. 13·3, 4·1.

Exercise 12 (p. 123)

2. Orig. distr., 16·9, 5·93; samples given, 16·05, 2·77.
3.

	Mean	S.D.
Orig. distr.	4·42	4·77
Samples of 5	4·81	2·00
Samples of 10	4·79	1·58

Means of samples should agree, but for grouping error.
S.D.s compare with $4·77/\sqrt{5}$ (=2·13) and $4·77/\sqrt{10}$ (=1·51).
4. S.E. for sample of 500 = 0·213; dev. = 1·78 × S.E.; prob. = 0·075.
5. (i) 0·30, (ii) 0·71.
6. No; prob. = 0·123.
7. 5·94 to 6·50; 5·85 to 6·59.
8. Yes (diff. $\simeq$ 2 × S.E.); the houses were not equal in numbers.
9. Yes (diff. $\simeq$ 2·5 × S.E.); the second list contained a number of Indian names.
10. No (diff. = 1·45 × S.E.).
11. Yes (diff. of means 0·02, S.E. 0·027).
12. A/M Not significant (diff. = 1·36 × S.E.).
    V_2 Significant (diff. = 2·1 × S.E.).
    T Not significant (diff. = 1·46 × S.E.).
13. (i) (a) 7·35, 1·61, (b) 5·11, 1·11.
    (ii) (a) Null hypothesis: they cannot distinguish black and white. Result highly significant (over 3 times S.D.).
        (b) Null hypothesis: they cannot distinguish the colours. Result not significant.
    (iii) (a) Excess proportion = 1·36 × S.D. Not convincing (probability of doing as well by chance = 0·087).
        (b) Not significant.
    (iv) S.D. for the whole group of 66 = 1·67.
        Diff. of mean scores = 0·90, which is more than twice the S.E., showing a significant difference.

Exercise 13 (p. 130)

1. 217·5, 1·5 per cent.
2. 0·7, 0·29.

3. 1·5 per cent, 0·0062.
4. 7·46, 0·195 m/sec.
5. 269, 7·25.
6. 2·236, 0·037.
7. 0·0058.

Exercise 14 (p. 136)

1. Yes (diff. $\simeq$ 2·4 × S.D.; they probably missed some common factors).
2. Yes (diff. less than S.E.); prob. = 0·365 (two-tail test).
3. No (diff. $\simeq$ S.E.).
4. (i) No (diff. less than S.E.), (ii) 0·052 (one-tail test).
5. (*a*) 35 to 41 per cent, (*b*) 34 to 42 per cent.
6. 4·2 to 11·8 per cent (2nd approx. 5·2 to 12·5 per cent). Prob. = 0·15; a further test needed.
7. £145·2 to £154·8 (no replacements).
8. No: the distribution is unsymmetrical and therefore not normal.

Index

Area scale 10, 56
   under normal curve 95, 96
Aristotle 1
Arithmetical probability paper 106
Assumed mean 16
Averages 5, 59

Bernoulli, Jakob 2
Bernoulli's Theorem 2, 75, 77
Binomial distribution 45, 71, 140
Bivariate distributions 29, 139

Cardan 2
Central limit theorem 117
Closest fit, line of 26
Coefficient of
   correlation 31, 139
   regression 28
   variation 128
Combinations 43
Confidence intervals 61, 120, 133
Continuous variables 9, 66, 139
Control charts 84, 119
Correlation 30, 139
Covariance 27
Cumulative frequency 13
Curve fitting 26, 103
Curve of errors 3, 91

Deciles 13
de Moivre 2, 99
Deviation
   mean 16
   quartile 12
   standard 16

Diagrams 6
Difference of independent variables 58
Difference of means 121

Errors
   normal curve of 3, 91
   observational 99, 127
Expectation 51
   of life 52
Expected value 56
   of sum and product 68

Fermat 2
Frequency
   curves 65
   density 65
   distributions 4, 8, 98, 138
   theory of probability 38

Galton 29
Gauss 3, 99
Gosset, W. S. 119
Graunt 1

Halley 2
Histogram 10

Independence, statistical 42
Independent
   events 41
   variates 57

Index

Keynes, J. M. 38
King, Gregory 2

Least squares 3, 26
Legendre 3

Mean 13, 56, 138
  deviation 16
  of a sample 59, 100, 140
Median 12
Mode 10
Mutually exclusive events 41

Normal curve 91
  equation of 109
  fitting 103
Normal distribution 91, 140
Normal probability paper 104
Null hypothesis 62, 118

Parameters 115
Pascal 2
Percentiles 13
Permutations 43
Petty, Sir W. 1
Poisson
  distribution 81, 140
  summation chart 87
Population 1, 56, 115
Probability 1, 38
  conditional 39
  density 67, 139
  distributions 45, 56, 139
  generating function 46

Quality control 84
Quartiles 12

Random sample 116
Random variable 8, 56
Regression 28, 29, 139
Repeated trials 45, 71

Sample, mean of 59, 100, 140
Sampling 115, 140
  of attributes 131
  simple 134
  without replacement 135
Scatter diagrams 29
Second moment 17
Semi-interquartile range 12, 24
Shepherd's adjustment 69
Significance tests 60
  levels of 118, 132
Skew distribution 11, 15
Small samples 119
Stakes, division of 2, 52
Standard deviation 16
  of binomial distribution 75
  of Poisson distribution 82
Standard error of mean 61, 117, 141
Statistics 1, 115
Stirling, James 3
'Student' 119
Sum of independent variables 58, 86, 111

t-distribution 119
Table of areas under normal curve 95
Table of random numbers 116
Tchebycheff's Theorem 77

Unbiased estimate of variance 120, 121, 141

Variable 8
  continuous 9
  discrete 8
  random 8, 56
Variance 16, 138
  of binomial distribution 75
  of Poisson distribution 82
  of sum and difference 58, 68
Variate 56, 86

Weighted average 6